Isoprenoid Biosynthesis in Higher Plants

Isoprenoid Biosynthesis in Higher Plants

Editor
S.K. Mangal

2007
Gene-Tech Books
New Delhi - 110 002

ISBN 81-89729-30-6

Published by : **GENE-TECH BOOKS**
4762-63/23, Ansari Road, Darya Ganj,
NEW DELHI - 110 002
Phone: 41562849
e-mail: genetechbooks@yahoo.co.in

Printed at : **Tarun Offset Printers**
Delhi

PRINTED IN INDIA

Preface

Isoprenoids are a large and diverse class of naturally occurring organic chemicals similar to terpenes, derived from five-carbon isoprene units assembled and modified in thousands of ways. Most are multicyclic structures which differ from one another not only in functional groups, but also in their basic carbon skeletons. Plant isoprenoids are extensively used for their aromatic qualities. They play a role in traditional herbal remedies and are under investigation for antibacterial, anti-neoplastic and other pharmaceutical effects. They contribute to the scent of eucalyptus, the flavours of cinnamon, cloves and ginger and the colour of yellow flowers.

Plants produce a greater array of structurally diverse isoprenoids than other organisms, and these are conveniently divided into classes of primary and secondary metabolites. Isoprenoids that are primary metabolites include sterols, carotenoids, growth regulators, and polyprenol substituents of dolichols, quinones and proteins. Isoprenoids classified as secondary metabolites include monoterpenes, sesqui-terpenes and diterpenes, and it is this class of isoprenoids that are more unique to plants. The topic of plant isoprenoids encompasses a rich and broad subject area, both of which are derived from a fascination in the diversity of isoprenoids found in nature and intense investigations into their biosynthesis.

This book focuses on the current understanding in the biosynthesis of isoprenoids in higher plants. It will be a unique source of information for all those involved with plants as chemical factories of fine chemicals and those involved with the quality of food and ornamental plants. It will be useful in teaching graduate courses in the field of biosynthesis in plants.

Editor

Contents

1

Mevalonate Pathway for Isoprenoid Biosynthesis

Isoprenoids represent the most diverse family of natural products. They are all formally derived from the branched C_5 skeleton of isoprene 1 (Fig. 1). Isopentenyl diphosphate 19 (IPP) and dimethylallyl diphosphate 20 (DMAPP) (Fig. 2) correspond both to the biological equivalents of isoprene. The elucidation of their biosynthesis resulted from the first applications of labelling with radioactive as well as stable isotopes and ended up in the discovery of the mevalonate pathway, which starts from acetate 12 activated as acetyl-coenzyme A 13 (Fig. 2). The early steps of isoprenoid biosynthesis was essentially performed with liver tissues and yeast for the biosynthesis of cholesterol and ergosterol (9, Fig. 1), and was later confirmed for the biosynthesis of plant sterols and triterpenes.

Mevalonate 16 was thus accepted as the universal precursor of all isoprenoids in all living organisms. Discrepancies with this assertion were, however, reported. 14C-Labeled mevalonate was poorly incorporated into the carotenoids (e.g., 7, Fig. 1) from plant chloroplasts. This was interpreted in terms of lack of permeability of chloroplasts toward mevalonate. In addition, incorporation yields of radio-labelled mevalonate into plant mono- and diterpenes were usually very low, at least as compared to those into sterols, which are synthesized in the cytoplasm.

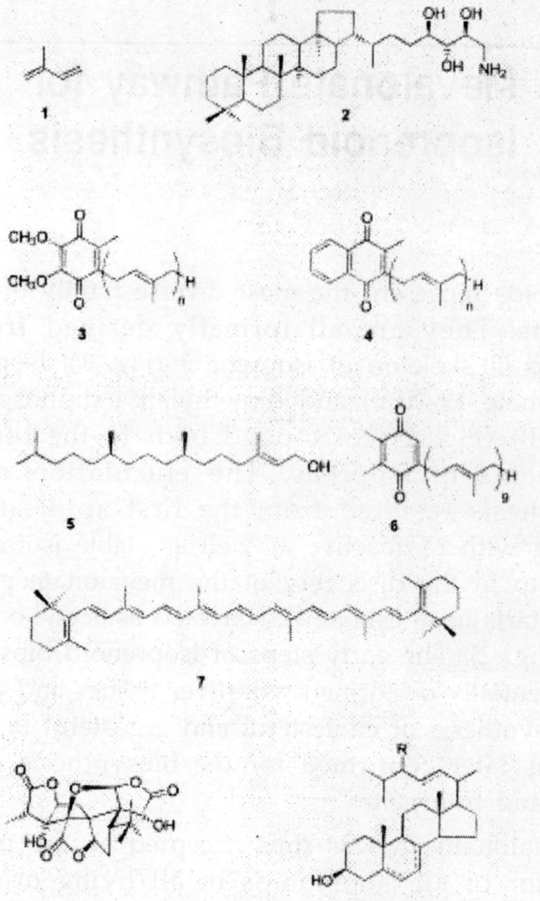

Figure 1. Isoprenoids investigated for the elucidation of the MEP pathway: isoprene **1**, *bacterial hopanoid* **2**, *ubiquinone* **3**, *menaquinone* **4**, *phytol* **5**, *plastoquinone* **6**, *β-carotene* **7**, *ginkgolide* **8**, *sterol* **9**.

Other data, which did not fit with the general occurrence of the mevalonate pathway, were obtained with mevinolin, a potent hypocholesterolemic agent, inhibiting the HMGCoA reductase, a key enzyme of the mevalonate pathway reducing hydroxymethylglutaryl-CoA 15 into mevalonic acid (MVA) 16.

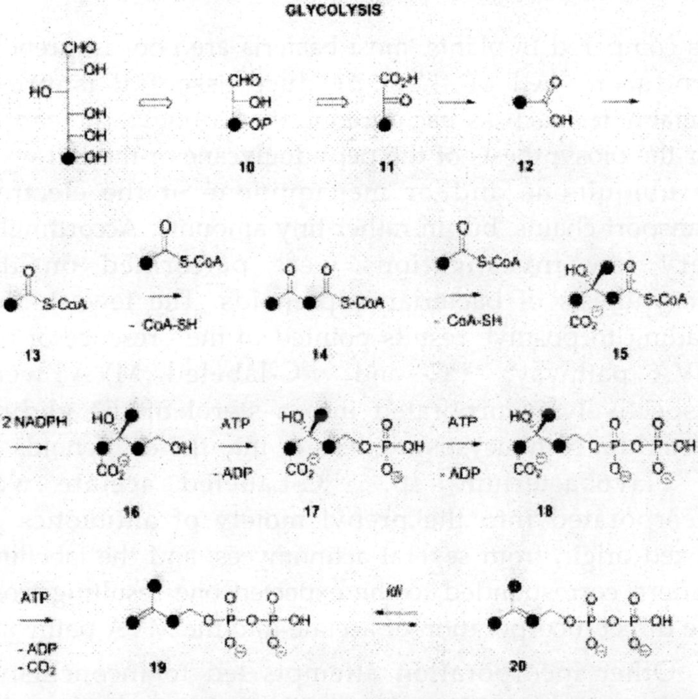

Figure 2. MVA pathway for isoprenoid biosynthesis with labeling pattern from [1-^{13}C]glucose metabolized via glycolysis.

Mevinolin strongly inhibits sterol biosynthesis in plant systems, but had no effect on carotenoid and phytol biosynthesis. Again, such negative results were interpreted in terms of impermeability of the chloroplast membrane toward the antibiotic. In addition, a cytoplasm-independent IPP biosynthesis was discovered in plant chloroplasts and chromoplasts. All these contradictory results did not really question the accepted dogma of the mevalonate pathway, and the identity of the biosynthetic route followed for the biosynthesis of chloroplast isoprenoids.

Mevalonate-Independent Methylerythritol Phosphate Pathway in Bacteria

As compared to plants, most bacteria are poor isoprenoid producers. All of them produce essential isoprene metabolites such as bactoprenol, the carbohydrate carrier for the biosynthesis of the peptidoglycane of the cell wall, or ubiquinone and/or menaquinone of the electron transport chains, but in rather tiny amounts. Accordingly, very few investigations were performed on the biosynthesis of bacterial isoprenoids. The few studies leading to positive results pointed to the presence of the MVA pathway. ^{14}C- and ^{13}C-labeled MVA were respectively incorporated into a sterol of the gliding bacterium Nannocystis exedens or into the carotenoids of a "Flavobacterium" sp. ^{13}C-Labeled acetate was incorporated into the prenyl moiety of antibiotics of mixed origin from several actinomyces, and the labelling pattern corresponded to the expected one resulting from the direct incorporation of acetate into the MVA pathway.

Other incorporation attempts led to inconclusive results, which could not be readily interpreted. Thus, the incorporation of ^{14}C-labeled acetate into the prenyl chain of ubiquinone from Escherichia coli resulted in a radioactivity distribution, which did not fit with the expected one from the MVA pathway, and led the authors to propose an alternative variant of this pathway via acetolactate. ^{13}C-Labeled acetate was not incorporated into this prenyl chain of ubiquinone, when E. coli was grown on a culture medium containing glucose and amino acid-derived carbon sources, whereas pyruvate labelled two contiguous carbon atoms derived from C_3 and C_5 of

DMAPP or IPP

Such results were indeed the premises for the presence of an alternative route for the formation of isoprene units. The discovery of a novel pathway for the early steps of isoprenoid biosynthesis was indeed a sideproduct of the investigations on the biosynthesis of bacterial triterpenoids of the hopane series (Fig. 1). A C_{35} skeleton, in which the triterpenic hopane moiety is linked by a carbon/carbon bond to a polyhydroxylated n-alkyl side chain, characterizes all major bacterial hopanoids. All bacteria do not produce hopanoids, but in the hopanoid-producing bacteria, such triterpenoids were shown to play a role of membrane stabilizers, much like sterol do in eukaryotic membranes.

Hopanoid concentration is usually of the same order of magnitude as the sterol concentration in eukaryotic cells. This makes them well suited for investigation by NMR spectroscopy after feeding of ^{13}C-labeled precursors. Feeding of ^{13}C-labeled acetate to the hopanoid producing bacteria Rhodopseudomonas palustris, Rhodopseudomonas acidophila, or Methylobacterium organophilum were performed with bacteria grown on a mineral medium only containing ^{13}C-labeled acetate as the only carbon and energy source. Such experimental conditions ensured the incorporation of the labeled precursor into the hopanoids.

As most biosynthetic pathways were known, a retrobiosynthetic analysis of the observed labeling patterns allowed reconstructing the reaction sequence involved in the formation of the hopanoids. Indeed, the experiments performed with [1-^{13}C]- and [2-^{13}C]acetate showed that the C5 side chain was derived from a D-pentose synthesized via the nonoxidative pentose phosphate pathway and linked to the triterpenic skeleton via its C5 carbon atom. The most interesting problem was, however, found in the isoprene moiety. The observed labeling pattern completely differed from the expected one from the MVA pathway. Interpretation was not obvious from the sole experiments performed with labeled acetate.

The clues for the elucidation of this metabolic pathway were obtained after incorporation of glucose isotopomers with ^{13}C labeling either at C1, C2, C3, C5, or C6 into the hopanoids of Zymomonas mobilis. Indeed, Z. mobilis only utilizes glucose as carbon source. This glucose is metabolized via the Entner-Doudoroff pathway. This bacterium does not possess a complete tricarboxylic acid cycle and does not convert pyruvate into glyceraldehyde phosphate because of the lack of an active enolase.

From the observed labeling patterns, it could be deduced that isoprene units in the hopanoids were derived from a C_2 subunit synthesized via pyruvate decarboxylation and from a C_3 subunit corresponding to a triose phosphate derivative (Fig. 3). These results were in accordance with the formerly described incorporations of ^{13}C-labeled acetate. The observed labeling pattern in the hopanoids from Rhodopseudomonas spp. or from M. organophilum simply resulted from the incorporation of acetate into the glyoxylate cycle, which allows growth on acetate as sole carbon source, and into the tricarboxylic acid cycle, yielding finally phosphoenolpyruvate, the precursor of pyruvate and the triose phosphate derivatives.

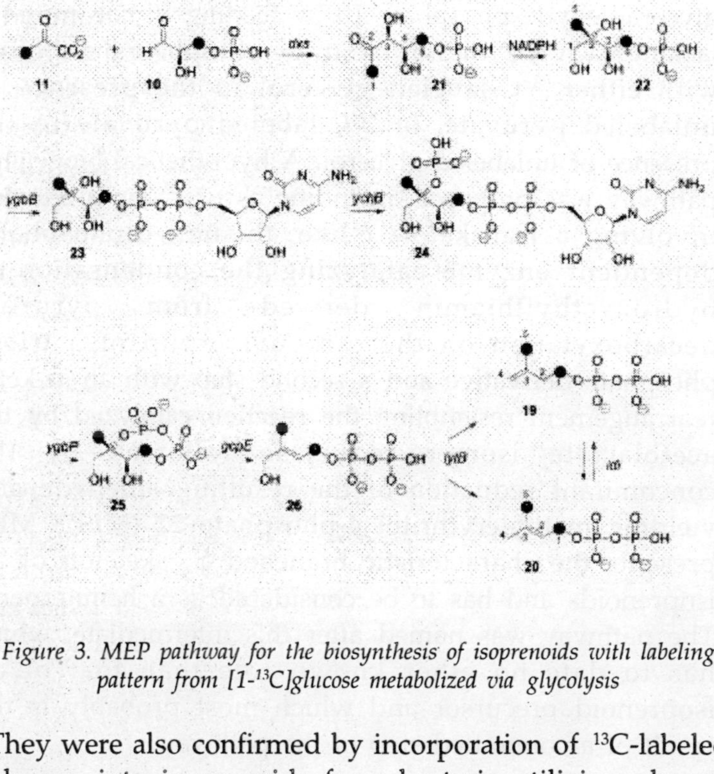

Figure 3. MEP pathway for the biosynthesis of isoprenoids with labeling pattern from [1-^{13}C]glucose metabolized via glycolysis

They were also confirmed by incorporation of ^{13}C-labeled glucose into isoprenoids from bacteria utilizing glucose via the Embden-Meyerhof-Parnas pathway and by studies in other isoprenoid series such as the ubiquitous ubiquinones and menaquinones. Incorporation of doubly labeled [4,5-$^{13}C_2$]glucose into the hopanoids and ubiquinone of Methylobacterium fujisawaense and of uniformly labeled [U-$^{13}C_6$]glucose into the hopanoids from Z. mobilis afforded a definitive proof for the formation of the isoprene units via a C_2 and a C_3 subunit and shed light on a rearrangement reaction allowing the insertion of the C_2 subunit derived from pyruvate 11 between two carbon atoms from the triose phosphate-derived C_3 subunit.

This triose phosphate was identified as glyceraldehyde phosphate 10 by feeding E. coli mutants lacking each enzymes of the triose phosphate metabolism with either ^{13}C-labeled glycerol in the presence of unlabeled pyruvate, or ^{13}C-labeled pyruvate in the presence of unlabeled glycerol. A hypothetical biogenetic pathway was proposed, including as first step a reaction involving a transketolase-like, thiamin diphosphate-dependent enzyme catalyzing the condensation of hydroxyethylthiamin derived from pyruvate decarboxylation on the carbonyl group of a triose phosphate derivative and a second step with an α-ketol rearrangement resembling the reaction catalyzed by the acetolactate isomero-reductase followed by the concomitant reduction of the resulting aldehyde and yielding methylerythritol 4-phosphate 22 (MEP). MEP presents the characteristic branched C_5 skeleton of all isoprenoids and has to be considered as a hemiterpene. The pathway was named after this intermediate, which has to date no other known role than that of an isoprenoid precursor and which most probably is the result of a committed step of the pathway.

MEP Pathway in Choroplasts of Phototrophic Organisms

Once an alternative route for isoprenoid biosynthesis has been found in bacteria, it was tempting to check whether such a pathway is involved in the formation of plant isoprenoids such as carotenoids, phytol, mono- and diterpenes. Indeed, independent IPP formation was formerly pointed out in plant chloroplasts, and the presence of an alternative route would be consistent with the discrepancies with respect to the MVA pathway.

Labeling experiments using ^{13}C-labeled glucose isotopomers were performed on ginkgo embryos by Schwarz and Arigoni, independently of our own assays with bacteria, and afforded the first proof for the presence of the MEP pathway in plants. They resulted in labeling patterns of ginkgolides 8 (Fig. 1), belonging to a diterpenoid series, which could be interpreted in the frame of the same biogenetic pathway as that proposed for the formation of bacterial isoprenoids (Fig. 3). They also shed light on the dichotomy in the isoprenoid biosynthesis in plant cells: the mevalonate pathway is involved, as expected, in the sterol biosynthesis in the cytoplasm, whereas isoprenoids that are most likely plastid related are synthesized via the MEP pathway.

Incorporation of ^{13}C-labeled glucose into the isoprenoids from several higher plant systems afforded additional proofs for the presence of the MEP pathway in plants. The ubiquitous terpenoids of the photosynthetic apparatus (carotenoids 7, phytol 5 from chlorophyll, and the prenyl chain of plastoquinone 6) are also derived from the MEP pathway. Finally, labeling experiments, performed with ^{13}C-labeled glucose or with deuterium-labeled deoxyxylulose, showed the wide distribution of the MEP pathway. It is involved in the formation of all investigated hemi- (e.g., isoprene 2, 2-methylbut-3-en-2-ol, dimethylallyl and the rest linked to aromatic rings), mono- and diterpenoids and some sesquiterpenoids.

Identification of Intermediates

1-Deoxyxylulose 5-phosphate

From the origin of the carbon atoms in the isoprene units derived from the MEP pathway, as deduced from the ^{13}C glucose labeling experiments, a 1-deoxy-D-xylulose (DX) derivative 21 (Fig. 3) was a likely isoprenoid precursor. This pentulose was not an unknown natural product. It has been found in the fermentation broth of a Streptomyces sp. and was known as precursor for pyridoxol phosphate and thiamin diphosphate in E. coli. Efficient incorporation of deuterium-labeled DX isotopomers into the prenyl chain of ubiquinone and menaquinone of wild-type E. coli represented the first proof for the role of a DX derivative as isoprenoid precursor. Similar experiments performed with other systems from bacteria or from plants confirmed this conclusion.

The gene of the 1-deoxyxylulose-5-phosphate (DXP) synthase (dxs or ispA) was rapidly identified in E. coli and in peppermint, and later in many other organisms, owing to its expected similarity with the transketolase genes. As predicted, the enzyme is thiamin diphosphate-dependent and utilizes GAP 10 and pyruvate 11 as substrates to yield DXP 21.

Derivatives of 2-C-Methylerythritol 4-phosphate 2-C-Methyl-D-erythritol (ME) is a natural product found in several higher plants. Its 2,4-cyclodiphosphate 25 (Fig. 3) is accumulated in large amounts by some gram-positive bacteria in oxidative stress conditions induced by one-electron transfer inhibitors such as benzylviologen. According to its structure, this tetrol fits well into the biogenetic scheme. It was, therefore, tempting to check whether ME is an isoprenoid precursor. Incorporation of ^{13}C-labeled glucose showed that the carbon skeleton of ME from Corynebacterium ammoniagenes is synthesized via the same reaction sequence as the isoprene units of the menaquinones.

Deuterium-labeled ME was incorporated by wild-type E. coli into the prenyl chain of ubiquinone and menaquinone with a localization of the labeling in accord with the hypothetical biogenetic scheme. This allowed the search for *E. coli* mutants auxotrophic for ME. Analysis of such mutants was followed by the discovery of the DXP isomero-reductase gene (*dxr or ispB*). This enzyme catalyzes, as predicted, the acid-catalyzed α-ketol rearrangement of DXP as well as the reduced nicotinamide adenine dinucleotide phosphate (NADPH)-dependent concomitant reduction of the resulting aldehyde, yielding MEP 22 as reaction product. The enzyme requires a divalent cation (Co^{2+}, Mg^{2+}, or Mn^{2+}) for catalysis.

The next three steps correspond to an activation of MEP, yielding three novel intermediates. Incubation of 3H-labeled MEP in the presence of nucleotide triphosphate with crude cell-free systems from E. coli resulted in the formation of a novel metabolite. The purification and identification of this novel radiolabeled metabolite was, however, difficult. The search in databases for genes encoding enzymes using a polyol phosphate (such as MEP) and a nucleotide triphosphate as substrate revealed the homology of a gene encoding an enzyme catalyzing the formation of 5′-diphosphocytidyl ribitol from cytidine 5′-triphosphate and from ribitol 5-phosphate and of a gene of unknown function in *E. coli* (*ygbP or ispC*).

Cloning and overexpression of the E. coli ygbP gene resulted in the identification of an enzyme using MEP and cytidine triphosphate as substrates, yielding 4-diphosphocytidyl ME 23 as reaction product. Whereas all other genes of the MEP pathway are scattered in the E. coli genome, this gene forms a small operon with the two next ones. This feature facilitated their identification. The enzyme encoded by the *ychB* (or *ispD*) gene catalyzes the adenosine 5'-triphosphate (ATP)-dependent phosphorylation of the tertiary hydroxy group of 4-diphosphocytidyl ME yielding 4-diphosphocytidyl ME 2-phosphate 24, and that encoded by *ychB* (or *ispE*) the formation of ME cyclodiphosphate 25 from the former intermediate with elimination of cytidine 5′-monophosphate (CMP).

The three novel intermediates were incorporated into carotenoids by red pepper chromoplasts, suggesting that they corresponded to intermediates of the MEP pathway. The latter three genes were also identified by another approach. In an E. coli strain capable of utilizing MVA after introduction of the genes permitting its conversion into IPP (MVA kinase, phosphomevalonate 17 kinase, and diphosphomevalonate 18 decarboxylase), random mutations, which were rescued by the addition of MVA to the culture medium allowed the identification of the *ygbP, ychB,* and *ygbB* genes and indicated that they are essential genes of the MEP pathway.

Independent IPP and DMAPP Biosynthesis

The formation of allylic alcohol diphosphates such as IPP 19 and DMAPP 20 from an ME derivative (a tetrol) is not an obvious process. In order to get more insight into the steps downstream of ME cyclodiphosphate, more information was required. Knowledge on the fate of the hydrogen atoms in the pathway was expected to afford useful clues for the further elucidation of the MEP pathway. Labeling experiments using deuterated DX isotopomers and a wild-type E. coli or deuterated ME isotopomers and an E. coli mutant with no functional deoxyxylulose phosphate synthase dxs gene showed that no apparent changes occurred at the level of the carbon atoms corresponding to C1, C4, and C5 of IPP or DMAPP. All corresponding deuteriums from DX or from ME were preserved in the prenyl chains of ubiquinone 3 and menaquinone 4.

Feeding of wild-type E. coli with [4-^{2}H]DX 27 yielded a surprising labeling pattern of the prenyl chains (Fig. 4).

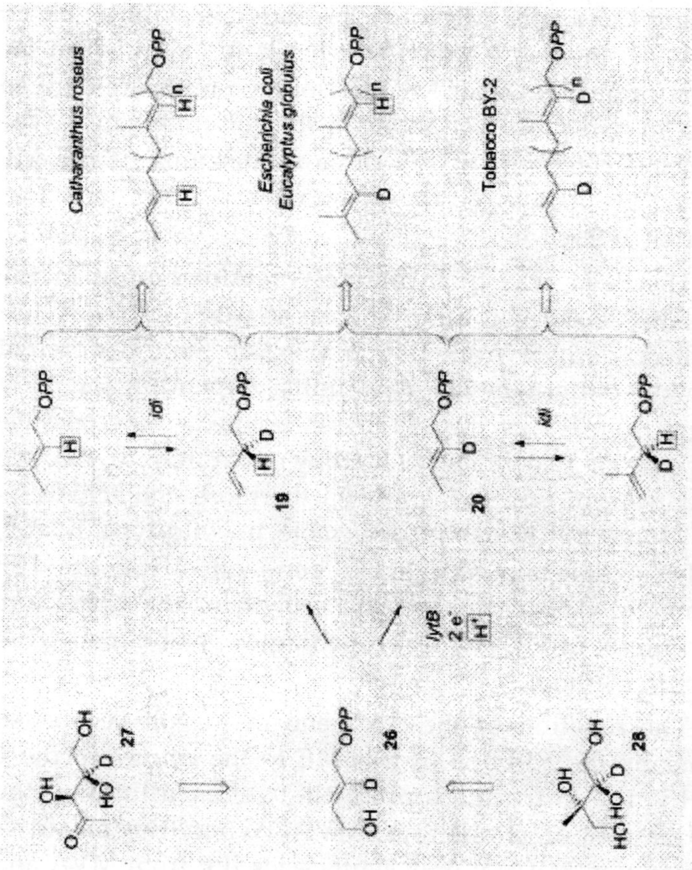

Figure 4. Independent IPP and DMAPP biosynthesis in the MEP pathway. Incubation of [4-²H]-1-deoxy-D-xylulose 27 or [3-²H]-2-C-methyl-D-erythritol 28.

The terminal isoprene units derived from DMAPP 20 and serving as starter for the elongation of the prenyl chain by the prenyl transferase showed considerable deuterium retention (up to 70 %), whereas all those derived from IPP 19 had no trace of deuterium. This was confirmed by feeding [3,5,5,5-$^{2}H_{4}$]ME to an E. coli strain with a deleted dxs gene. In such a strain, which has to be grown in the presence of free DX or ME, all isoprene units are derived from labeled precursor added to the culture medium. The expected retention of the three deuteriums of the ME methyl group was observed, and their signal served as internal reference.

The 3-^{2}H was quantitatively retained only in the DMAPP 20 derived unit, whereas it was completely lost in those derived from IPP 19 (Fig. 4). Such a different labeling pattern in isoprene units, depending on their origin from IPP or from DMAPP, pointed to the possible presence of two branches leading separately to the two precursors of isoprene units. Furthermore, the activity of IPP isomerase is barely detectable in wild-type E. coli, and the deletion of the *idi* gene encoding the IPP isomerase does not affect the growth of the bacterium. Definitive proof of the branching was obtained by a genetic method.

The deletion of the dxr gene in E. coli is rescued either by the addition of free ME to the culture medium or by the introduction of the genes allowing the utilization of exogenous MVA. When the strain is growing on MVA, it only synthesizes IPP from MVA and has to rely on a functional protein capable of isomerizing IPP into DMAPP. If in addition the idi gene is deleted, the novel strain does not grow anymore on MVA, indicating that there is no other way in E. coli for converting IPP into DMAPP than the reaction catalyzed by the protein encoded by the idi gene. It grew, however, normally on ME. This shows that IPP and DMAPP can be independently synthesized from MEP derivative by two separate branches: an IPP branch characterized by deuterium loss from [4-2H]DX or from [3-2H]ME and a DMAPP branch characterized by deuterium retention.

The fate of the 4-H hydrogen of DX can thus be utilized for the detection of a possible branching of the MEP pathway in plants (Fig. 4). Incubation of a Catharanthus roseus cell culture with [1-^{13}C, 4-^{2}H]DX resulted in ^{13}C labeling of all isoprene units with no trace of deuterium, suggesting that no branching occurred, at least in the utilized culture conditions. Feeding of [4-^{2}H]DX to eucalyptus twiglets was followed by a slight (0.9 %) but significant deuterium incorporation into the monoterpene cineol, but only in the DMAPP-derived moiety, with nearly no deuterium found in the isoprene unit derived from IPP. Finally, incorporation of [4-^{2}H]DX into the plastid isoprenoids from a tobacco BY-2 cell culture resulted in the labeling of all isoprene units of phytoene and plastoquinone, whatever their origin from IPP or from DMAPP was, with the same isotope abundance (ca. 15 %).

These experiments revealed the presence of the branching in plant cells. In C. roseus, only the IPP branch was detected. In eucalyptus, like in E. coli, both IPP and DMAPP branches were found. In tobacco, the DMAPP branch significantly contributed to the formation of all isoprene units. Definitive interpretation of the labeling patterns was only possible once more insight into the reduction reactions involved in the last two steps has been obtained.

From Methylerythritol Cyclodiphosphate to IPP and DMAPP

Two additional genes, gcpE (or ispG) and *lytB* (or *ispH*), were always found accompanying the formerly identified genes of the MEP pathway. Deletion of these genes was lethal for E. coli, but rescued by the insertion of the three genes allowing the utilization of exogenous mevalonate, indicating that they were implied in this biosynthetic route. An E. coli strain with a deleted *gcpE* gene and capable of utilizing mevalonate accumulated methylerythritol cyclodiphosphate 25, suggesting that this intermediate is the substrate of the corresponding *GcpE* enzyme. Crude cell-free systems from an E. coli strain overexpressing the biosynthesis of GcpE were capable of converting ME cyclodiphosphate 25 into (E)-4-hydroxy-2-methylbut-2-enyl diphosphate 25 (HMBPP).

This protein was later shown to possess an oxygen-sensitive [4Fe-4S] cluster involved in a reductive process corresponding to two successive one-electron transfers. Regeneration of the reducing system could be performed either by flavodoxin/flavodoxin reductase/NADPH or by deazaflavin via the semiquinone radical. The accumulation of HMBPP 26 in a *lytB*-deficient E. coli mutant strongly suggested that HMBPP is the substrate of the *LytB* protein. A different approach also shed light on the role of the *gcpE* gene.

An E.coli strain overexpressing the gene of the xylulose kinase, an enzyme capable of phosphorylating free DX, and all genes encoding enzymes located upstream of the *GcpE* protein in the MEP pathway accumulated ME cyclodiphosphate upon incubation with uniformly ^{13}C-labeled DX. This experimental procedure allowed the detection of intermediates by directly running the ^{13}C NMR spectra on crude cell-free systems. Additional overexpression of the *GcpE* gene was followed by the accumulation of HMBPP 26 and of *LytB* resulted in the simultaneous formation of IPP 19 and DMAPP 20.

This represented the first indication that the *LytB* protein was responsible for the conversion of HMBPP 26 into IPP 19 and DMAPP 20 and that the reaction catalyzed by this protein corresponds to the branching previously described. Finally incubation of 3H-labeled HMBPP with chromoplast from Capsicum annuum and Narcissus pseudonarcissus led to the formation of radiolabeled carotenoids, confirming the role of HMBPP 26 as intermediate in the MEP pathway.

2

Carotenoid Biosynthesis

Carotenoids are synthesized in bacteria, algae, fungi, and green plants. In addition to very few bacterial carotenoids with 30, 45, or 50 carbon atoms, C40-carotenoids represent the majority of the more than 600 known structures. Especially bacterial carotenoids are most diverse. Hydroxy groups at the ionone ring may be glycosilated or carry a glycoside fatty acid ester moiety.

Furthermore, carotenoids with aromatic rings or acyclic structures with different polyene chains and typically 1-methoxy groups can be found (Fig. 1A). Typical fungal carotenoids possess 4-keto groups, may be monocyclic, or possess 13 conjugated double bonds (Fig. 1B). 3-Hydroxy a- and β– as well as 5,6-epoxy β–carotene derivatives (Fig. 1C) are abundant in chloroplast of some algal groups and green plants.

Reaction Sequence

The mevalonate pathway to prenyl pyrophosphates. In fungi, carotenoids are derived via the mevalonate biosynthetic pathway. The reaction sequence is shown in Fig. 2A. The metabolite mevalonate is formed from

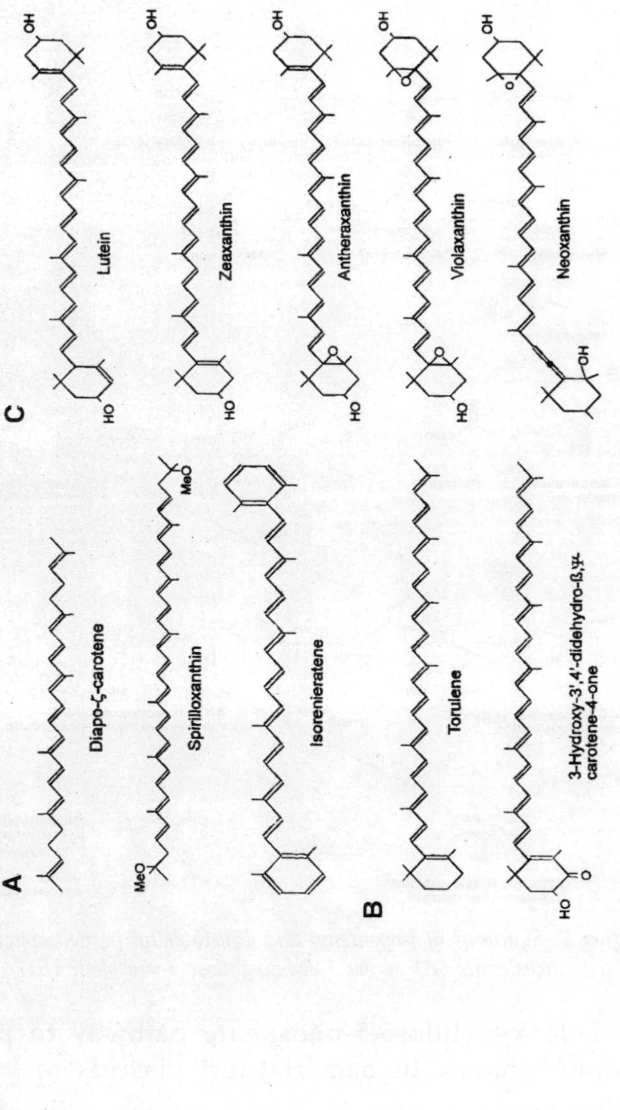

Figure 1. Carotenoids typical for bacteria (A), fungi (B), and plants including green algae (C).

acetyl-CoA via hydroxymethylglutaryl-CoA. The subsequent steps involve two kinase reactions yielding mevalonic acid pyrophosphate which then is decarboxylated to isopentenyl pyrophosphate (IPP)[2]. IPP is the building unit for the synthesis of terpenoids.

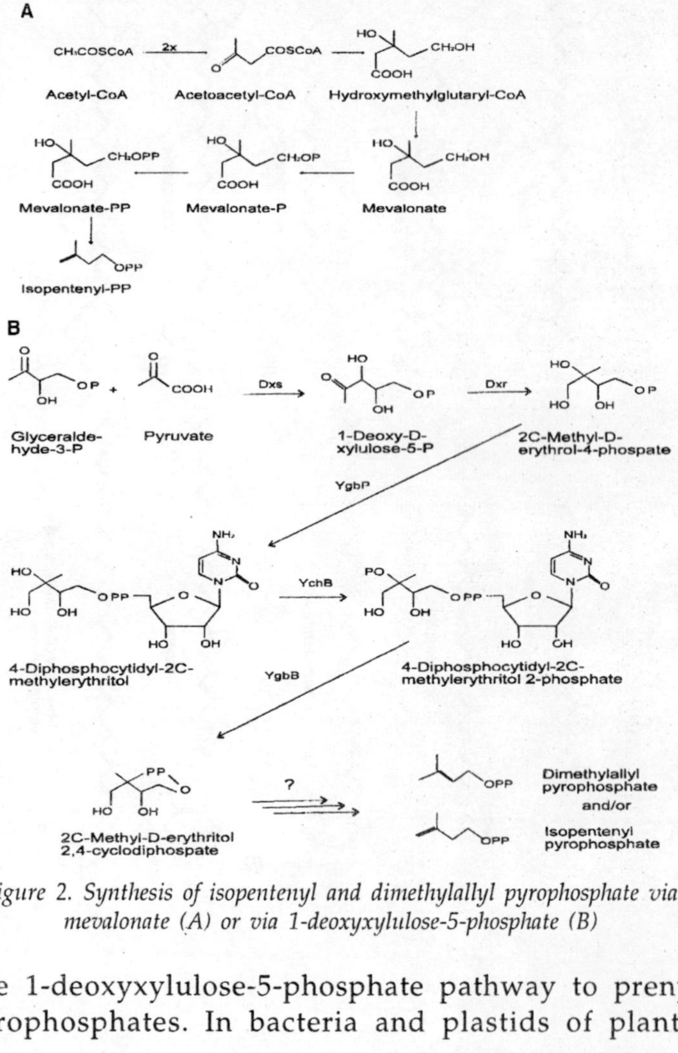

Figure 2. Synthesis of isopentenyl and dimethylallyl pyrophosphate via mevalonate (A) or via 1-deoxyxylulose-5-phosphate (B)

The 1-deoxyxylulose-5-phosphate pathway to prenyl pyrophosphates. In bacteria and plastids of plants, formation of prenyl pyrophosphates which are the precursors of carotenoids proceed via an alternative pathway. This novel pathway is referred to either as the 1-deoxyxylulose-5-phosphate pathway since this is the first intermediate or as the 2-C-methyl-D-erythritol-4-phosphate pathway since this is the first product which is converted to prenyl pyrophosphate exclusively. Not all of the reaction steps are elucidated to date.

Initially, a C_2-unit from pyruvate is condensed to glyceraldehyde by a thiamin-dependent 1-deoxyxylulose-5-phosphate synthase (Fig. 2B). Then, the product is converted to 2-Cmethyl-D-erythritol by 1-deoxyxylulose-5-phosphate reductoisomerase. The following reactions involve the formation of 4-diphosphocytidyl-2-C-methylerythritol in a CTP-dependent reaction and phosphorylation at position 2 by ATP. Subsequently, CMP is released and a 2-C-methyl-D-erythritol 2,4-cyclodiphosphate is formed.

This product may be the branching point for independent routes to IPP and dimethylallyl pyrophosphate (DMAPP). For example, cleavage of phosphate from C-2 may result either in a 1-keto or 3-keto pyrophospate intermediate. However, the final steps to IPP and DMAPP are still open. Evidence is accumulating that the pathways of prenyl pyrophosphate synthesis differ in prokaryotes and eukaryotes. In Synechocystis the formation of IPP and DMAPP must be parallel as indicated in Fig. 3A since no IPP isomerase gene exists in this species and IPP isomerase activity is absent.

The situation is similar to Escherichia coli where a parallel synthesis of IPP and DMAPP has been established. Nevertheless, an IPP isomerase is present in E. coli and its overexpression leads to increased formation of carotenoids in the corresponding transformant. Since

IPP isomerase is present in plant plastids and its activity is the limiting step of carotenoid synthesis in etioplasts, it may be concluded that IPP is the major or only product of the 1-deoxyxylulose-5-phosphate pathway. Then DMAPP should originate from isomerization of IPP. Since the possibility of parallel formation of DMAPP cannot be excluded yet, this reaction is indicated in Fig. 3A with a question mark.

Formation and Interconversion

From prenyl pyrophosphates of different chain lengths, specific routes branch off into various terpenoid end products. The prenyl pyrophosphates are formed by different prenyl transferases after isomerisation of IPP to DMAPP by successive 19-4 condensations with IPP molecules. The carbon bond is formed between the C-4 of IPP and the C-1 of the allylic co-substrate. In the case of geranylgeranyl pyrophosphate (GGPP) synthesis (Fig. 3B) this is dimethylallyl pyrophosphate in higher plants or farnesyl pyrophosphate in many bacteria. In all prenyl transferase reactions the mechanism of this head-to-tail joining is basically the same.

Phytoene synthesis is the first committed step in C40-carotenoid biosynthesis (Fig. 4). It proceeds as a head-to-head condensation of two molecules of GGPP forming a cyclopropylcarbonyl between carbon 1 of the first and carbons 2 and 3 of the second GGPP molecule yielding prephytoene pyrophosphate as an intermediate. In plants, 15-cis phytoene is then formed by elimination of the pyrophposphate group and stereospecific proton abstraction. Subsequently, four dehydrogenation steps are carried out in the conversion of phytoene to lycopene which is the plant carotene with the highest number of conjugated double bonds. Two completely unrelated types of phytoene desaturases exist.

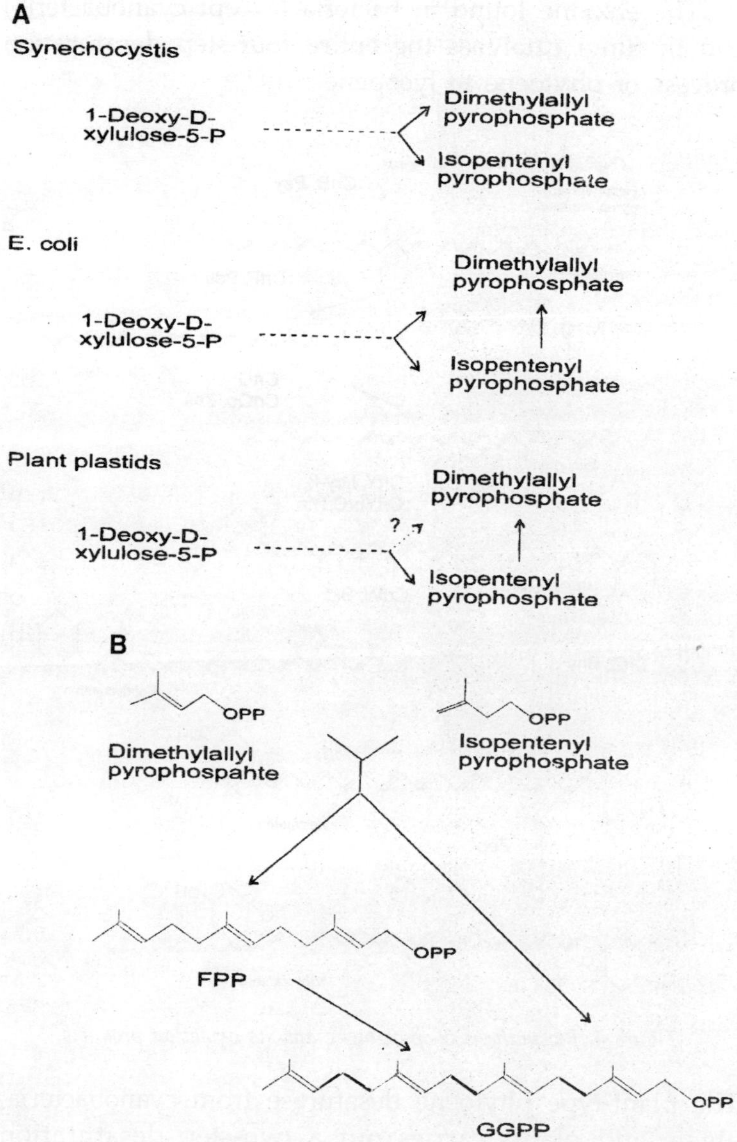

Figure 3. Sequential or parallel formation of isopentenyl and dimethylallyl pyrophosphate in different organisms (A) and their conversion to geranylgeranyl pyrophosphate (B).

The enzyme found in bacteria (except cyanobacteria) and in fungi catalyzes the entire four-step desaturation process of phytoene to lycopene.

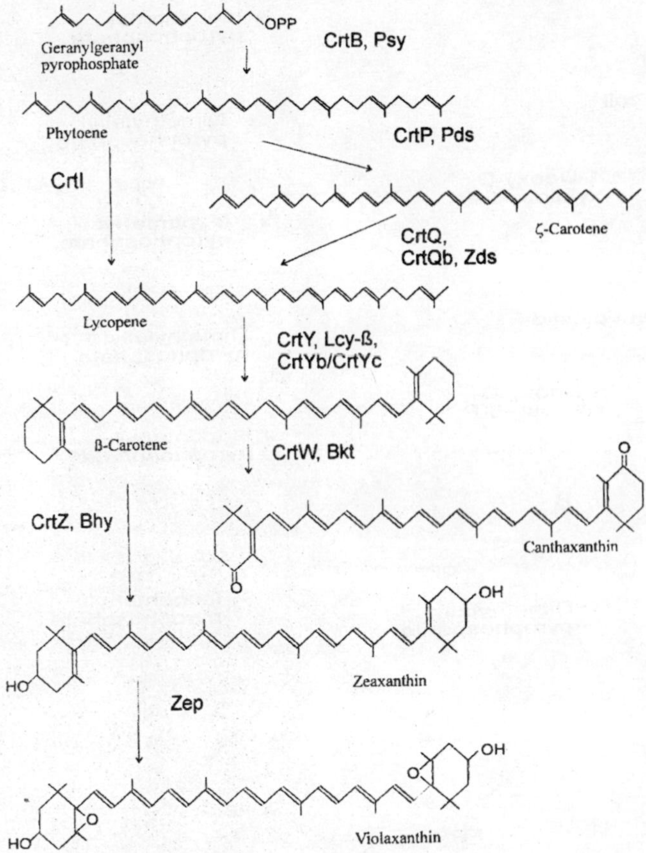

Figure 4. Biosynthesis of β-carotene and its oxidation products

The plant-type phytoene desaturase from cyanobacteria, algae, and plants carries out a two-step desaturation reaction with different ζ–carotene stereoisomers as reaction products. Lycopene is the regular substrate for cyclization reactions. Two types of ionone rings are formed: the β–ring as an end group of both sides of β-carotene and the ε–ring at one side of a-carotene in addition to a β–ring.

The mechanisms of cyclization which involve proton attack at C-2 and C-29 of lycopene and addition of the C-1 carbocation to the C-5,6 double bond is substantially supported by D2O-labeling ex-periments followed by determination of the position of the deuteriated carbon at the β– or ε–ionone ring. The resulting carbonium ion intermediate is stabilized by loss of a proton either from C-1 or C-4 to yield a β– or ε–ring, respectively. Formation of the β– or ε–ring is under different gene control. Consequently, we can assume that formation of β-carotene is catalyzed by a single enzyme, lycopene cyclase β.

In case of a-carotene, two different enzymes, lycopene cyclase b and lycopene cyclase ε, are involved. Genetic complementation analysis demonstrated that the ε-cyclase forms only one ring whereas the β–cyclase introduces two rings into lycopene or an additional ring into δ-carotene which is a lycopene molecule with one ε-end group. Xanthophylls are enzymatically formed oxidation products of a- and β–carotene (Fig. 1). Typical oxo groups are hydroxy at C-3 of the β– and ε–ionone ring, epoxy at the 5,6 position of the β–ionone ring, and keto groups at position 4.

Enzymic Reactions and Biosynthesis

In the past 10 years, the biochemistry of carotenogenesis and especially the cloning of carotenogenic genes has made considerable progress. To date, many genes for different carotenogenic reactions are known. Detailed information on their cloning, their structure, and their function can be found in several reviews. The availability of these genes was a great help in the characterization of enzymes of carotenogenic pathways and their regulation. In return, biochemical insights facilitated the cloning of novel carotenogenic genes.

Enzymes involved in carotenoid biosynthesis are all membrane-associated or integrated into membranes. This feature together with their sensitivity against detergents which have to be used for solubilization and their very low abundance makes the isolation and purification rather difficult. This is the major reason why only a few have been purified from plant tissue. Many others have been overexpressed in a heterologous host in a way that highpressure cell breaking resulted in a soluble and enzymatically active form.

Furthermore, gene expression allows the extension of the coding region by an affinity group, e.g., a 6xHis tag which facilitates purification of the expressed enzymes considerably. Phytoene synthase. This enzyme is encoded by the closely related bacterial crtB and eukaryotic psy genes. The crtB gene encoding phytoene synthase from the eubacterium Erwinia uredovora was overexpressed in E. coli. The phytoene formed in the enzymatic reaction was present in both a 15-cis and all-trans isomeric configuration.

The essential cofactors required were ATP in combinations with either *Mn21* or *Mg21*. Phytoene synthesis was inhibited by phosphate ions and squalestatin. In tomato, enzymatic studies of wild-type and mutant fruit in the green and ripening stage demonstrated the existence of two different phytoene synthases, *Psy-1,* encoded by the formerly called pTOM5 cDNA, and *Psy-2,* which both mediate the formation of 15-cis phytoene. These enzymes are expressed organelle-specific. *Psy-2* predominates in green tissue and is responsible for carotenoid synthesis in chloroplasts whereas *Psy-1* is expressed during fruit ripening and contributes to carotenoid tomato fruit chromoplasts. The chromoplast enzyme has been isolated from ripe Capsicum annuum fruit. In addition, the phytoene synthase from Narcissus pseudonarcissus has been obtained after heterologous expression of the cDNA.

The most detailed characterization was carried out with a purified preparation from green tomato fruit of *Psy-2* which provides information not only about its biochemical properties but also on its interaction with other carotenogenic enzymes. This work revealed the existence of a very close association between phytoene synthase and other proteins, particularly those involved in the synthesis of other terpenoids. This association may be important for the effective functioning of the enzymes in vivo, by enabling the channeling of biosynthetic precursors.

Phytoene desaturase. Different phytoene desaturases are known. They differ in the number of desaturation steps and in their structures. Among them is the bacterial/fungal type encoded by crtI and the cyanobacterial/algal/plant type encoded by *crtP* or *pds*. Bacterial type phytoene desaturases. Phytoene desaturase from E. uredovora was expressed in E. coli. Purification was achieved after solubilization of the protein sequestered in inclusion bodies by urea treatment. Reactivation was possible by removal of the denaturant and dilution of the sample in the presence of dithiothreitol. It could be shown that FAD was involved as cofactor in the desaturation reaction of this bacterial type of phytoene desaturase.

Plant type phytoene desaturase. In cyanobacteria and plants desaturation of phytoene to lycopene proceeds in two steps which are catalyzed by two individual enzymes, phytoene desaturase and ζ–carotene desaturase.

Phytoene desaturase converts phytoene to ζ–carotene with phytofluene as an intermediate. The enyzme from Synechococcus was expressed and puri-fied. The resulting 53-kDa membrane protein could be reactivated after lipid replenishment. Inhibition was observed by several bleaching herbicides. Substrates in addition to phytoene were phytofluene and 1,2-epoxy phytoene which were converted to ζ–carotene and the corresponding 1,2-epoxyde.

The reaction was stimulated by NAD, NADP, and oxygen. The phytoene desaturase from C. annuum chromoplasts has been solubilized by detergent and purified to a polypeptide of 52 kDa with a reported FAD prosthetic group. The cofactor for the desaturation reaction remains obscure. The phytoene desaturase from N. pseudonarcissus chromoplasts is dependent on oxidized quinones as hydrogen acceptor. Regeneration of reduced quinone occurs with oxidation with molecular oxygen by an oxidoreductase, a homodimeric enzyme with a molecular mass of the subunit of 23 kDa.

In leaves, a plastidic alternative oxidase involved in phytoene desaturation exists as exemplified for Arabidopsis thaliana. A model has been proposed on the participation of this alternative oxidase and of the photosynthetic electron transport chain in different developmental and physiological stages.

ζ–Carotene desaturases. Since the phytoene desaturases encode by crtP and pds synthesize ζ–carotene as end product, a second enzyme must exist for the formation of lycopene. This ζ–carotene desaturase catalyzes the final two desaturation steps to lycopene by introducing two double bonds at positions 7,8 of ζ–carotene and 79,89 of neurosporene. From cyanobacteria two structurally unrelated gens for ζ–carotene desaturase were cloned.

One *crtQ* (formerly called zds) from Anabaena is related to the bacterial crtI gene, whereas the second *crtQb* (also called *crtQ-2*) from Synechocystis is quite similar to crtP. This gene is very closely related to the plant zds gene. ζ–Carotene desaturase *CrtQ*. The ζ–carotene desaturase *CrtQ* from Anabaena was expressed in E. coli, solubilized from the membranes by Chaps, and puri-fied to homogeneity. Substrates for this enzyme apart from ζ–carotene are those carotenes which partially resemble the latter, like neurosporene and β–zeacarotene yielding lycopene and g-carotene, respectively, as reaction products. Also cis isomers like pro- ζ–carotene were converted to the corresponding products.

ζ–Carotene desaturase CrtQ-2/Zds. The desaturase Zds from the plant C. annuum was expressed in E. coli, purified, and biochemically characterized. Product analysis showed that different lycopene isomers are formed including substantial amounts of the all-trans form together with 7,79,9,99-tetracis prolycopene via the corresponding neurosporene isomers. Application of different geometric isomers as substrates revealed that the ζ–carotene desaturase has no preference for certain isomers and also revealed the nature of the isomers formed during catalysis. The hydrogen acceptors for the dehydrogenation of ζ–carotene and neurosporene are lipophilic quinones including plastoquinone.

Lycopene cyclases. Two completely unrelated types exist in bacteria, fungi, and plants. One encodes the classical monomeric bacterial β–cyclase gene, crtY, which may be an ancestor of crtL, the gene for a β–cyclase in cyanobacteria. From crtL the plant β– and ε–cyclase genes developed. In gram positive bacteria two genes crtYc and crtYd are present. They encode two proteins which have to interact as a heterodimer for lycopene β–cyclization.

From this type of lycopene cyclase genes the fungal lycopene cyclase/ phytoene synthase fusion gene crtYB evolved.

Lycopene β–cyclase CrtY from E. uredovora has been purified to homogeneity in an active state after expression in E. coli. In addition to its two-step reaction in which both sides of the lycopene molecule are cyclized to β–ionone rings with the monocyclic g-carotene as an intermediate, neurosporene as well as 1-hydroxylycopene were cyclized to β–zeacarotene and hydroxyg-carotene, respectively. The cofactors involved in the reaction were either NADH or NADPH. Also the lycopene β–cyclase Lcy-b was expressed in E. coli, purified, and subjected to enzyme kinetic studies.

Complementation studies with the β–lycopene cyclase from C. annuum demonstrated that this enzyme, like the β–cyclase from E. uredovora, is not specific for the cyclization of the ψ-end group as found in lycopene and at one end of neurosporene, but also is able to convert the 7,8-dihydro-ψ-ends to 7,8-dihydrob–end groups yielding 7,8,-dihydro-β–carotene in addition to β–zeacarotene from neurosporene. In *C. annuum* fruit another enzyme with homology to lycopene β–cyclase was found.

This enzyme also exhibits lycopene β–cyclase activity but its major catalytic activity is the formation of capsanthin and capsorubin from violaxanthin or antheraxanthin, respectively. This capsanthin-capsorubin synthase (CCS) is the only enzyme from a biosynthetic pathway leading to secondary carotenoids which has been purified. β–Carotene hydroxylase. Bacterial β–carotene hydroxylases, CrtZ, were characterized with respect to substrate specificity and reaction mechanism after expression in E. coli.

The reactions depend on oxygen and are stimmulated by 2-oxoglutarate, ascorbate, and Fe^{2+}, which is typical for a dioxygenase reaction. Similar mechanistic properties were also found for bacterial ketolases encoded by crtW. Of the two distinct plant carotene hydroxylases which introduce a hydroxy group at position 3 of either an ε–ionone or a β–ionone end group, only the β–carotene hydroxylase has been purified. The enzyme originated from expression of a *C. annuum* cDNA in E. coli.

The conversion of β–carotene via β–cryptoxanthin to zeaxanthin was strictly dependent on the presence of molecular oxygen. In addition, the catalytic activity was strictly dependent upon reduced ferredoxin which could not be replaced by NADPH. Treatment of puri-fied hydroxylase with the iron chelators o-phenanthroline and 8-hydroxyquinoline strongly inhibit activity, indicating the presence of iron as a functional group.

Zeaxanthin epoxidase. This enzyme introduces 5,6-epoxy groups into a 3-hydroxy-β–ionone end group. It was expressed in E. coli from a *C. annuum* cDNA. The recombinant protein acts specifically on β–cryptoxanthin, zeaxanthin, and antheraxanthin. The epoxidation reaction involved oxygen and reduced ferredoxin as reductant.

Neoxanthin synthase. A cDNA encoding neoxanthin synthase has been cloned very recently from tomato (45). Based on mechanistic similarities between the neoxanthin synthase and capsanthin-capsorubin synthase reactions forming a transient carbocation during the opening of the cyclohexyl 5,6,-epoxides, oligonucleotides against CCS cDNA sequences were used to screen a neoxanthin synthase cDNA from a tomato library. This cDNA which encodes a 56-kDa plastidtargeted protein was expressed in E. coli. The resulting transformant was functional in converting violaxanthin into neoxanthin.

Genetic Manipulation

The availability of a large number of carotenogenic genes makes it possible to modify and engineer the carotenoid biosynthetic pathways in microorganisms and plants. These molecular biology approaches aimed for carotenoid production in heterologous microorganisms, plants resistant to bleaching herbicides, and plants with enhanced nutritional value.

Carotenoid Production in Noncarotenogenic Hosts

Fungi. Carotenoids have been successfully synthesized in noncarotenogenic fungi. The first attempts have been made with Saccharomyces cerevisiae which was transformed with the carotenogenic genes from E. uredovora for the production of lycopene. Since then, Candida utilis has been developed systematically as a production host for lycopene, β–carotene, and astaxanthin. The exogenous carotenoid biosynthesis genes originating from the eubacterium *E. uredovora* and the marine bacterium Agrobacterium aurantiacum were modified based on the codon usage of C. *utilis* and expressed under the control of constitutive promoters derived from this host. The resultant yeast strains accumulated lycopene, β–carotene, and astaxanthin by directing the carbon flow into carotenoid production at the expense of ergosterol biosynthesis.

Carotenoid formation could be further increased by metabolic engineering of the mevalonate pathway. Expression of the catalytic domain of hydroxymethylglutaryl-CoA reductase gene together with the disruption of the endogenous erg9 gene encoding squalene synthase yielded lycopene concentrations of 7.8 mg/g dry weight.

E. coli. E. coli is a very convenient host for heterologous carotenoid production. Most of the carotenogenic genes from bacteria, fungi, and higher plants can be functionally expressed in this bacterium. Furthermore, plasmids belonging to different incompatibility groups with different antibiotic resistance markers are available. They can all be introduced simultaneously and combined in E. coli for carotenoid synthesis, allowing combinations of individual genes.

By combination of genes from different organisms with different carotenoid biosynthetic branches, novel carotenoids not found in any other pathway can be synthesized. The potential of E. coli as a carotenoid production system has been reviewed recently. Carotenoid production of E. coli can be increased by metabolic engineering of the supply of prenyl pyrophosphates as a precursor for carotenoid production.

Transformation with the genes for overexpression of 1-deoxy-D-xylulose 5-phosphate synthase, 1-deoxy-D-xylulose-5-phosphate reductoisomerase, and isopentenyl pyrophosphate isomerase stimulated carotenogenesis up to 3.5-fold to a final yield of 1.5 mg/g dry weight (50).

Engineering of resistance to bleaching herbicides. The typical bleaching herbicides directly interact with phytoene desaturase of plants as the target enzyme. Resistance to these herbicides was engineered in tobacco by introducing the crtI gene from E. uredovora. Enzyme kinetic studies of the encoded phytoene desaturase showed that this enzyme is naturally resistant to bleaching herbicides.

Since the endogenous tobacco phytoene desaturase is located in the thylakoid membrane, the 59 region was fused to the sequence for a transit peptide and put under the control of the cauliflower mosaic virus 35S promoter.

After transformation of tobacco, expression and processing of the corresponding protein could be demonstrated. The resulting tobacco plants were resistant not only to bleaching herbicides interfering with phytoene desaturase but also to inhibitors which affect ζ–carotene desaturase.

Carotenoids in tomato fruit. In several crops the carotenoid content was manipulated by transfer of foreign carotenogenic genes. During lycopene deposition in tomato fruit ripening, phytoene synthase is the controlling enzyme. Therefore, this enzyme was the first target for the genetic manipulation of the carotenoid composition of tomato fruit. The constitutive high level expression of *Psy1* in tomato plants resulted in carotenoid rich seed coats, cotyledons, and hypocotyls. However, hemizygous and homozygous plants were reduced in stature due to changes in the giberellic acid and abscisic acid contents.

Manipulation of the desaturation activity in tomato was achieved with the same plasmid containing the crtI gene from E. uredovora which was used for the generation of herbicide resistant tobacco. Expression of the gene did not change the amount of total carotenoids in transgenic tomato fruit but resulted in a threefold increase of the β–carotene content.

Rapeseed. Overexpression of a phytoene synthase gene from E. uredovora extended with a plastid targeting sequence under the seed-specific napin promoter was a successful attempt to increase the carotenoid content of mature rapeseed up to 50-fold. In the transformant the embryos were bright orange compared to the green embryos in the control. In the trans-genic seeds, carotenoid concentrations of more than 1 mg/g fresh weight accumulated, consisting mainly of aand β–

carotene. Concurrent to the enhanced carotenoid production, the chlorophyll and a-tocopherol contents decreased in the transformant.

Rice endosperm. Rice was transformed with a plasmid containing plant phytoene synthase gene and a bacterial phytoene desaturase gene which should mediate the synthesis of lycopene from GGPP. Both reading frames were extended with transit sequences for the targeting of the endosperm plastids. They were under the control of the endosperm-specific glutelin and the constitutive cauliflower mosaic virus 35S promoter, respectively.

Surprisingly, the accumulated carotenoids were lutein and zeaxanthin a- and β–carotene instead of lycopene. Thus, it was assumed that the lycopene β– and ε–cyclases and the β– and ε–hydroxylases must be induced in the transformant or constitutively expressed. Cotransformation with a lycopene β–cyclase containing plasmid increased the β–carotene content. Even then, the total amount of 1.6 mg/g carotenoid in the rice endosperm is only 1 to 2% of the carotenoid concentration in the transgenic rapeseed.

3

Photoregulation of Cartenoid Biosynthesis

Since development is based on specific alterations of metabolism, such as the formation of enzymes and their products, photoregulation of biosyntheses is also a common phenomenon; the most striking example is the obligatory light-dependence of chlorophyll formation in higher plants. Besides this effect on chlorophyll formation, photoregulation of carotenoid biosynthesis has also been reported not only in higher plants, but also in algae, fungi and non-photosynthetic bacteria. Why do many plants produce carotenoids only as a response to illumination? has this photoregulation any ecological meaning? During the development it may be very economical for the organism to form structures and compounds only when they are needed.

The main functions of carotenoids in plants, i.e. their role in photosynthesis and protection of the organism against the potentially harmful effects of irradiation, are related only to growth of the plant in the light. Consequently, the bulk of carotenoids is needed when e.g. the seedling emerges from the ground or fungal mycelia grow into illuminated areas. Therefore, photoregulation acts quite clearly as a "saving mechanism" to avoid waste of both material and energy and may provide the organism with an ecological advantage.

Difficulties in Photoregulation

The main problem in the photoregulation of carotenoid biosynthesis is—as in the photoregulation of other processes—how can the organism translate the physical signal "light" into a biochemical signal and finally into physiological and biochemical events? The sequence of reactions involved, beginning with the reception of the light signal and ending up with the final response—biosynthesis of carotenoids—is usually termed the "mechanism" of the photoregulation. For convenience we may divide the regulatory chain into several steps.

Photo reception—Signal transduction—Level of regulation—Synthesis

Although it is difficult to separate the reaction completely, some general print.-ciples of the different steps shall be described briefly.

Photoreceptor pigments. The usual approach to discovering the photoreceptor that is responsible for a certain response, is a comparison of the action spectrum of the response with the absorption spectra of putative pigments. However, besides action spectroscopy further investigations using other methods are necessary for an unequivocal identification of the acting photoreceptor. At least three of the photore-ceptors, discussed as being responsible for light-mediated carotenogenesis—chlorophyll, phytochrome and a designated "cryptochrome"—are also very important for other photo responses in plants.

Signal transduction. In signal transduction twc different types of regulation are important which also have implications for the level of regulation. In the first situation accumulation of carotenoids proceeds only during illumination, that is the permanent presence of the inducing factor "light" is essential for the response or, in the seconds light is needed only as a trigger whereas the response takes place also in darkness. The latter case is the so-called classical "induction" mechanic

Level of photoregulation. Considering the biochemical background of all known photo-regulations of biosynthetic processes, in particular the molecular machinery for the formation of enzymes, in principle there are four possible levels for the action of light, for all of them well documented examples analogous to photoregulation of carotenogenesis are known:

— A direct transformation of a light-absorbing compound—a certain carotenoid -which then facilitates subsequent biosynthetic steps;

— photoinduced changes in the activity of carotenogenic enzymes already presented;

— de novo synthesis of carotenogenic enzymes from messenger-RNA already present by regulation of trans.-lation or &.

— de novo synthesis of the enzymes via regulation of transcription, theat is de novo synthesis of the appropriate messenger-.RNA's.

The mechanism for 3 groups of plants—algae, higher plants and fungi—it will be demonstrated that they represent different types of mechanism with respect to photo- receptors, signal transduction and level of photoregulation.

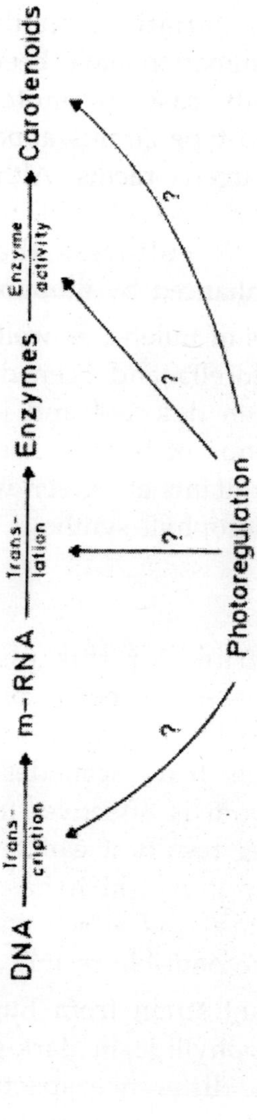

Figure 1. Possible levels for the action of light in photoregulation of carotenoid biosynthesis

Alge

Algae form intact chioroplasts and the usual plastid carotenoids when grown in the dark; this indicates that carotenoid biosynthesis in algae exhibits no light-requirement, although variations in the carotenoid pattern induced by illumination have been reported for some species. The only case of photoregulation of caroteno- genesis in wild-type strains appears to be that of the phytoflagellate Eugena racilis. Although cultures have the capacity to synthesise all caro- tenoids that are present in light-grown cultures, the amount of carotenoids is greatly enhanced by illumination.

However, in mutants of Eulena, as well as in mutants of the green algae Chlorella and Scenedesmus, strong photoregulation has been detected and investigated in more detail by the groups of Schiff, Claes, Britton and Senger. Some of these mutants are green but others have lost the capacity for chlorophyll synthesis.

Photo Receptors

From an action spectrum for this photoregulation determined for Eug1ena it has been concluded that a protochiorophyllide is the acting photoreceptor. In chlorophyll-free mutants from Scenedesmus, only the blue part of the spectrum is effective; from the action spectrum and additional results it can be assumed that the blue-light photoreceptor which has recently been designated "cryptochrome" and which will be discussed later in more detail is responsible in this case.

In a bleached mutant strain from Euglena which is deficient in protochlorophyllide in dark-grown cultures blue light minus dark difference spectra indicated a photo-isomerisation of *cis-* to *trans-ζ-carotene*. Since additional results have shown that no compound other than cis-ζ–carotene is the photoreceptor responsible for its own isomerisation one might speculate that this is the rate-limiting step for the subsequent biosynthesis of more unsaturated carotenes and xanthophylls.

Level of Photo Reulation

The formation of acyclic carotenoids are presents that means such carotenoids are formed independently of light, and illumination triggers the formation of cyclases and hydroxylases. However, kinetics of the changes in pigment composition during illumination, some of which are shown in Fig. 2.

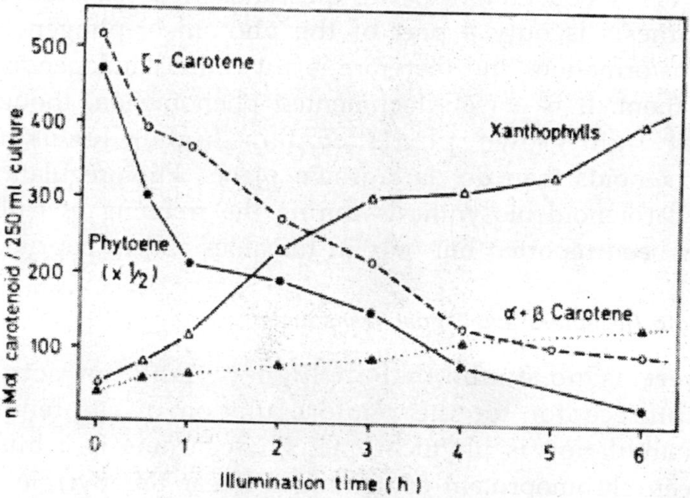

Figure 2. Pigment changes following illumination of a dark—grown culture of Scenedesmus obliguus mutant PG 1

This figure shows a rapid fall in phytoene and -carotene concentrations immediately after the onset of light, that is without any lag-period. Furthermore, they also show a concomitant and immediate increase of 13-carotene and xanthophylls.

Higher Plants

Seedlings of higher plants grown in the dark have some capacity for carotenoid biosynthesis but only angiosperms show a light-dependent stimulation of carotenoid production during the development of the young plant. In investigations on photo regulation in angiosperm seedlings one has to consider that the complete transformation of proplastids or etioplasts to photosynthetically active chloroplasts, including the synthesis of chlorophyll, is completely light-dependent.

As a consequence, the stimulation of carotenoid synthesis is only a part of the photom or-phogenetic transformations and therefore is not entirely independent of them. It is a well-documented phenomenon, though that light-grown plants contain higher levels of carotenoids than do dark-grown plants. Photoregulation of carotenoid biosynthesis during the ripening of fruits has been reported but only in tomatoes and in paprika.

Photo Receptors and Signal Transduction

There is no doubt that in higher plants an acting photoreceptor for the photoregulation of carotenoid accumulation is phytochrome. Phytochrome is a blue-green chromoprotein containing a linear tetrapyrrole as chromosphere. It is specific for plants and exists in two interconvertible light-absorbing forms—a red-light and a far-red-light form. Since phytochrome is an important photoreceptor in photo morpho-genesis of higher plants its structure, photochemical reactions and its mode of action have been intensely discussed.

Goodwin and his co-workers were the first to find that total carotenoid formation in seedlings is stimulated by a brief illumination with red light and that this effect can be nullified by a subsequent exposure to far-red-light; further investigations conferred these results. Although an action spectrum determined for carotenoid accumulation in etiolated wheat leaves indicate proto chlorophyll rather than phytochrome as the photoreceptor, further characteristics of this photo induction closely resemble those of other photoregulations mediated by phytochrome. Therefore, by analogy, the mechanism in this case, that is pulse illumination, is that of a classical induction reversion phytochroine reaction in which the far-red absorbing form of phyto-chrome produced on exposure to red light acts as an inducer.

A different type of mechanism appears to be involved when dark-grown seedlings are exposed to continuous far-red light. Only prolonged illumination in-creases the rate of carotenoid accumulation after a lag-period; soon after the light is turned off, pigment production is reduced to the dark rate. Such so-called "high irradiance reactions" are characteristic of other developmental responses of seedlings which are continuously illuminated. Action spectra of such responses exhibit, in addition to a peak in the far-red region effectiveness of light in the blue and near UV region of the spectrum. It is dis-cussed, in the context of whether high irradiance reactions are mediated by two photoreceptors, phytochro:ne and the blue light photoreceptor, cryptochrome.

Level of Photo Reulation

The shape of the accumulation kinetics during continuous

far-red light shows that carotenoid formation is dependent on continuous illumination. Obviously, this type of photoregulation is different from that of an "induction" mechanism mediated by phytochrome via pulse illumination or by blue light as described later for fungi. Consequently, for the level of photoregulation it might be considered that as in the case of the algae some kind of "photo modulation" is involved the mechanism of which has yet to be elucidated.

As long as illumination operates only via phytochrome without a concomitant biosynthesis of chlorophyll (red-light pulses, continuous far-red light), accuinu-lation of carotenoids is enhanced to a moderate extent only. Furthermore, the carotenoid pattern virtually unchanged compared to that in complete darkness. This effect seems to be due to a stimulation of early step(s) in the biosynthetic pathway. If, however, seedlings are illuminated with white or continuous red-light leading to the synthesis of large amounts of chlorophyll and to the transformation of etioplasts to chloroplasts carotenoids production is increased drastically. A distinct change in the carotenoid pattern was found concomitantly.

Fungi

Photoregulation of carotenoid biosynthesis in fungi has been detected in only a few species: Neurospora crassa, Phycomyces blakesleeanus, Fu—sarium aquaeductuum and Verticillium aaricinum. The organisms synthesise carotenoids characteristic of the particular species. Wild—type Phycomyces produces β—carotene as the main pigment, intermediates are found in only trace amounts.

In contrast, in Fusarium, Neurospora and Verticilliup these intermediates which can be arranged according to the desaturation sequence first proposed by the late Prof.

Porter are accumulated to some extent. Whereas in Verticilliurn carotenoids accumulate only during illumination the kinetics of photo-induced accumulation in Fusarium (Fig. 3) and also in Neuro-Spora show different characteristics: Following a lag-perio after a brief illumination the amount of pigments increases rapidly for a certain time, and thereafter net pigment synthesis ceases; only continuous illumination leads to a prolonged accumulation.

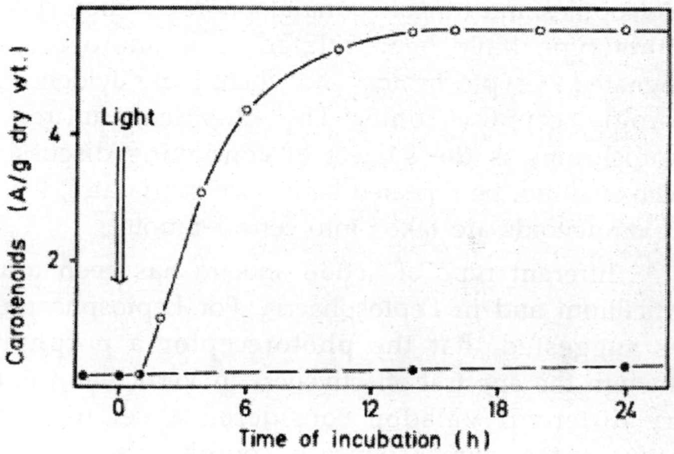

Figure 3. Accumulation of carotenoids in roung mycelia of Fusarium aguaeductuum in the dark (closed circles — •) and following a brief illumination (open circles — o). The arrow indicates the period of brief illumination.

The data gathered from experiments with brief illumination periods clearly show that the dark reactions are a strict consequence of photo reactions indicating that in the sequence of events (Fig. 1) a very early "point of no return" exists. This type of regulation exhibits all features of a "classical" induction mechanism.

Photo Receptors

Zalokar was the first to determine an action spectrum of carotenogenesis in Neurospora. This spectrum have following characteristic features: a maximum at 370—380 nm and three peaks or at least shoulders between 400 and 500 nm. Light of wavelengths greater than appr. 520 nm is ineffective. The shapes of the action spectra for carotenoid biosynthesis resemble those for a variety of developmental and movement responses in plants. The most prominent among these is the phototropic reaction of coleoptiles and fungal sporangiophores. Action spectra of this type have been related to a photoreceptor designated "Cryptochrome", but there is no evidence for a single cryptochrome. The chemical nature of cryptochrome is the subject of continuing discussion, which shall not be repeated here. Two candidates, flavins and carotenoids are taken into consideration.

A different type of action spectra has been found Verticillium and in Leptosphaeria. For Leptosphaeria, it was suggested that the photoreceptor a porphyrin. Although the epectral effectiveness in Verticillium is not very different, Valadon considered a yet unknown photoreceptor pigment since no porphyrins have been identified in the nrcelium of this fungus. In Verticillium, Valadon and his co-workers found, in addition to the effect of UV-blue light some effect also by light in the red region of the spectrum.

The photo-reaction shows the typical induction by red light and subsequent reversion by the far-red. However, since additional spectroscopic data and further results cannot fully be explained in terms of phytochrome transformations, the involvement of phytochrome needs further clarification.

Signal Transduction

For optimal photo-induction the presence of oxygen during illumination is essential. ilowever, a minor induction is possible in the absence of oxygen. Moreover, when mycelia were illuminated in an atmosphere of nitrogen with saturating fluences the photoreceptor system can be reactivated by oxygen in the dark. We therefore con-cluded that oxygen is not involved directly in the photochemical reaction of the photoreceptor but rather functions as an electron acceptor in keeping the photo-receptor—we consider a flavoprotein—in a proper state of oxidation.

The reducing agent dithionite applied to mycelia immediately after illumination prevents photo-induction. On the other hand incubation of the mycelia with buffered hydrogen per-oxide solution in the dark can mimic, to a certain extent,the effect of light in inducing carotenoid synthesis. Biosynthesis of carotenoids is also induced when mycelia are incubated with methylene blue and then illuminated with red light which is not effective in photoinduction via the natural photoreceptor.

Photo-induction by red light was found to be triggered only in the presence of photo-dynamically active redox dyes. From these results and on consideration of results from other blue-light mediated phenomena, we have drawn a hypothetical scheme illustrating the events during photoinduction via the cryptochrome photoreceptor.

Level of photorepulation

In contrast to the phytochrome-mediated regulation of carotenoid accumulation in seedlings in mycelia of Fusarium and Neurospora pigment production after a first illumination (Fig 3), and renewed synthesis after a second light treatment, starts only after a lag-phase. The occurrence of a lag-phase always points to time-consuming events intercalated between photoinduction and the final events. Earlier results, using several fungal species, have shown that photo-induced carotenoid accumulation is completely blocked by cycloheximide—a potent inhibitor of protein synthesis—when applied prior to or immediately after illu-niination of the mycelia. Iron these results it has been concluded that light induces a de novo synthesis of carotenogenic enzymes which are absent or present only in minor amounts in dark grown cultures.

An inhibitor of transcription in fungi—Distamycin—and an antimetabolite of adenosine in ItNA-synthesis—Tubericidin—blocked light-induced pigment formation in Fusarium. Recent investigations show that the inhibition of carotenoid synthesis is reduced gradually with the time elapsing between illumination and application of the inhibitor. Moreover, the time course of the decrease of inhibitory power is different for the inhibitors of messenger-ILNA synthesis and protein synthesis. This might indicate that synthesis of carotenogenic enzymes is preceded by the synthesis of the messenger-ItNA's coding for the synthesis of these enzymes.

The synthesis of the carotenogenic enzymes is regulated at the level of transcription. However, the involvement of light-induced gene expression remains speculative unless specific photoregulated messenger-UNAs have been identified. As a first step towards such direct evidence, Schrott some years ago found that, shortly after photoinduction, there is an increase in the relative amount of polyadenylated 1tNA being synthesized. Mitzka-Schnabel reports in her paper on *carotenogenic enzymes in Neurospora*, the presence of polypeptides in which radioactivity after in vivo-labeling with 35S-methionine was increased in response to photoinduction. Therefore, we started to extract messenger-RNAs and to investigate their capacity for translation in an in vitro-translation system from rabbit reticulocytes.

Cellular poly(A)-1tNA extracted from dark grown and illuminated mycelia of Neurospora and Fusarium revealed high translational capacity in vitro reflected by the stimulation of 35S-methionine incorporation. The amount of poly(A)-RNA is increased in illuminated mycelia in both fungal species; this might indicate a higher biosynthetic activity during carotenogenesis. The products of the in vitro translation were then separated by two-dimensional el-ectrophoresis on polyacrylamid gels. Fluorographs show that some in-vitro translated polypeptides increased in response to photoinducation.

4

Lipid Biosynthesis

Lipids are a class of hydrocarbon-containing organic compounds essential for the structure and function of living cells. Lipids are characterised by being water-insoluble and soluble in nonpolar organic solvents. Although the term *lipid* is often used as a synonym for fat, the latter is in fact a subgroup of lipids called triglycerides.

Vegetable Oils

The oils produced by plants provide a major source of calories for human nutrition. In addition, they provide the most concentrated source of energy of any food, supply essential fatty acids (which are precursors for prostaglandins and other similar hormones), are carriers for fat soluble vitamins, and serve to make foods more palatable and to the feeling of satiety after eating. Consumption of fats differs drastically in different parts of the world. As individuals and societies become more affluent, the proportion of calories from fats and oils in the diet tends to increase.

Country-specific analysis of data for 1988-1990 found a range of 7-46% for the % of dietary calories from fat. A total of 19 countries fell below the minimum recommendation of 15% dietary energy supply from fat, the majority of these being in sub-Saharan Africa and the remainder in South Asia. In contrast, 24 countries were above the maximum recommendation of 35%, the majority of these countries being in North America and Western Europe. An additional trend relating to fat/oil consumption has been the transition from consumption of animal to vegetable sources for fat. For example, in 1920 animal sources were responsible for 81% of fat calories consumed in the U.S., whereas by 2000, vegetable oils provided over 54% of fat/oil calories.

Oil Crops

The major crops which produce approximately three-fourths of vegetable oil for human consumption are soy, palm, canola and sunflower. However, a much larger number of oil crops are consumed because of their special properties, or are important in local economies. For example, sesame, olive and peanut oils are often used for the special flavours they impart to foods whereas cottonseed provides a major edible oil in Australia and China and other regions where cotton is a major crop and the seeds are a low-cost by-product of the fibre.

For large-scale use in baking and frying, the food industry chooses commodity vegetable oils largely based on price, which is related to the location of production and whether the oil is a by-product. For example, in the U.S. soybean oil is often considered a by-product of the soy meal industry, and its 80% "market share" dominates the vegetable oil market, making it the largest single source of calories in most U.S. diets. In contrast, palm and rapeseed are crops grown almost exclusively for their oil and are particularly prevalent in Asia and Europe, respectively, near their site of production.

Although hundreds of different fatty acid structures can occur in seeds of different plant species, only four fatty acids; palmitic, oleic, linoleic and linolenic make up 95% or more of the composition of the major edible vegetable oils. However, the relative proportions of these fatty acids influence both the physical properties and the nutritional impact of the oils.

For example, the saturated fatty acid content has a major impact on the melting point of the oil or fat. To produce shortening and margarines, in most cases vegetable oils are hydrogenated which increases their saturated fatty acid content, provides more solid physical properties and also leads to side reactions which yield trans fatty acids. Saturated and trans fatty acids also influence the nutritional properties of the oil, leading to increased cholesterol levels.

Although human tissues have the capacity for fatty acid synthesis, when fatty acids are abundant in the diet this biosynthetic capacity is suppressed. Consequently, most of the fatty acids present in human tissues are derived from the diet rather than from de novo synthesis. However, after consumption, dietary fatty acids can be acted on by selective enzymes which target certain structures to specific membrane components (or to other lipids) and they can also be modified in their structure by further desaturation, elongation or beta-oxidation (chain-shortening) reactions.

Nevertheless, despite these selections and modifications, the lipid composition of humans is strongly influenced by the fatty acid composition present in the diets. Because different human populations have very different diets, and because these same populations have very different causes of mortality, some members of the medical research community conclude that dietary fatty acids have a major impact on diseases. For decades, an emphasis on risks associated with saturated fatty acids and its relation to cholesterol metabolism has encouraged the consumption of more vegetable oil rather than animal fats. More recently, concern has increased about the role of the trans fatty acid isomers which are produced during the hydrogenation of vegetable oils and may have negative consequences. Also, in recent years evidence has accumulated that the balance of omega-3 and omega-6 unsaturated fatty acids in diets influences a wide range of human physiological responses including coronary heart disease (CHD).

The dominance of plant oils with high omega-6 18:2 in many diets (e.g. U.S.) has led to omega-6/omega-3 consumption ratios near 10:1 in many western diets whereas populations which consume ratios near 1:1 have strikingly lower incidence of CHD. One explanation offered for the strikingly different CHD mortality associated with high omega-6/omega-3 consumption ratios is the different fates of these structures in the biosynthetic pathways leading to prostaglandins and other eicosenoids. The omega-6 fatty acids are precursors to arachidonic acid which in turn produces prostaglandins and thromboxanes which promote blood clotting. In contrast, the omega-3 fatty acids are poor substrates for prostaglandin biosynthesis and act as competitive inhibitors of the arachidonic pathway.

The above issues regarding the potential influence of dietary fatty acids on human health have stimulated interest in use of genetic engineering to design healthier plant oils. Such efforts have led to some impressive technical successes in reducing the saturated fatty acid and omega-6 content of edible oils and in the future may provide abundant and inexpensive alternatives to fish oil for omega-3 fatty acids.

Genomic Approaches in Plant Fatty Acid Synthesis

In addition to their nutritional significance, vegetable oils are a major agricultural commodity, with worldwide production of 90 billion pounds, worth nearly $50 billion per year. This market size has attracted considerable interest to modify plant fatty acid production for both food and non-food uses. Major aims for molecular genetic improvement of plant oils have been to improve the total oil content of seeds, and to increase the content of specific target fatty acids in the oil to improve nutritional or functional properties and expand uses.

In several instances, a functional genomics approach has been successfully used to identify the genes responsible for the production of specific fatty acids. More recently, the development of genomics technology has opened doors for large-scale approaches aiming to uncover and understand the regulatory factors that govern synthesis and accumulation of fatty acids in tissues.

Lipid metabolism in the model plant Arabidopsis thaliana: a first draft of the metabolic maps and a catalog of the proteins/genes possibly involved. There is now a growing line of evidence that cellular metabolism is spatially organised and not just a collection of enzymes, substrates and effectors randomly dispersed in the aqueous phase of the cell. Evidences include enzyme arrays adsorbed to membranes or the cytoskeleton, multienzyme complexes and even metabolite pools that are not in free solution.

Moreover, enzymatic reactions can often be catalysed by several isoforms (about 2/3 of the genes in the model plant Arabidopsis thaliana belong to multiple gene families). The genes encoding different isoforms may correspond to a true functional redundancy, but could also bear specific information in their coding or regulatory sequences. Thus, different isoforms might be expressed in different tissues or organs at different stages, targeted to different organelles or subdomains within an organelle, regulated by different effectors and may interact specifically with other proteins or cell structures.

Therefore, only a set of specific isoforms may be relevant to a trait of interest and their identification can be a critical step in designing successful strategies of rational metabolic engineering. For the goal of engineering metabolic traits, the first step is to make a map of all possible reactions and a catalog of the proteins/genes that may be involved in each reaction. In recent years, this task has been greatly facilitated by the sequencing of entire genomes. Arabidopsis thaliana was a fortuitous choice for oilseed researchers to make such a catalog of reactions/ enzymes because this plant is:

— the best-studied model organism for plant biologists and the first plant to have its genome completely sequenced

— an oil seed plant (about 35% of lipids per seed dry weight)

— a species closely related to a major oil crop (canola).

A first survey of reactions, proteins and genes that may be involved in the metabolism of fatty acids and lipids derived from fatty acids (acyl-lipids) in Arabidopsis has been performed in our lab and reported before the complete sequencing of the Arabidopsis genome. This partial survey covers 65 reactions and compares Arabidopsis with rice. Based on this initial survey, we have recently reported a more complete catalog based on the fully sequenced genome, which covers most reactions of acyl-lipid metabolism in Arabidopsis.

The gene catalog is based on sequence similarity searches between the proteins predicted from the Arabidopsis genome and proteins from any organism that have been shown experimentally to be involved in a reaction of acyl-lipid metabolism. The candidate proteins have been subjected to multiple alignments and searches for conserved protein motifs. Over 600 proteins and encoding genes covering more than 120 reactions have been thus identified and classified according to predicted function, subcellular location and alternative splicing.

Candidate proteins with links to sequence databases and literature are available for each reaction by browsing either metabolic maps or a list of reactions of lipid metabolism sorted by organelles and main cellular functions. Only 13% of the 644 predicted proteins have been experimentally demonstrated to have lipid-related activity, so the vast majority of the genes represent uncharacterised proteins. Moreover, about 1/3 of these proteins were previously annotated as "unknown function" or with functions unrelated to acyl lipid metabolism. For each reaction, the catalogue gives a list of most probable candidate genes.

For example, regarding lipid synthesis, there are 4 main candidate genes for the endoplasmic 2-lysophosphatidate acyltransferase and 7 additional ones. This enzyme catalyzes the formation of phosphatidic acid, a key intermediate in membrane and storage lipid assembly. It is not known if phosphatidic acid and other precursors are synthesised by the same enzymes and at the same subcellular location for these two types of lipids. One of these acyltransferases might thus be specifically involved in oil accumulation in seeds.

In addition to the catalog, the database gives information about the distribution of ESTs in various organs for all genes of the catalog. A statistical analysis of EST abundance in organs is provided and lists of enzymes that are possibly up regulated in some organs is given. A tentative list of enzyme isoforms involved in storage lipid synthesis in seeds has for example been suggested. Clearly these lists of differentially expressed genes are only a first step in the search for isoforms of biotechnological interest and must be checked by experimental methods (northern blots, etc.).

It is expected that this Web-based database will provide a first common platform of information about reactions/enzymes of acyl-lipid metabolism in plants and that will facilitate functional studies in Arabidopsis and also in crop plants. This database can be particularly helpful in view of proteomics studies, where it will be important to recognise and distinguish different polypeptides from gene families for a particular reaction of acyl-lipid metabolism. It will also guide the search for knockout phenotypes and help in the design of gene knockout strategies as well as in the analysis of in vivo metabolic fluxes in mutants of lipid synthesis.

EST Sequencing Projects and Gene Discovery

ESTs (expressed sequence tags) are defined as single-pass DNA sequences from cDNA clones. Initial success was achieved in targeted gene discoveries using small scale EST sequencing projects where the tissues examined were enriched for tissue-specific transcripts. For example, genes responsible for production of ricinoleic acid and conjugated fatty acids were identified and cloned from ESTs prepared from oil producing tissues of castor bean, Momordica charantia and Impatiens balsamina. These fatty acids are industrially important for production of nylons, paints etc.

In recent years, due to greatly lower costs and improvements in DNA sequencing, larger scale sequencing projects have become common. The data generated from high throughput genomics can be used for annotating genes and understanding their spatial and temporal regulations. This information would be particularly helpful in understanding traits governed by multiple genes, especially quantitative traits. Furthermore, advanced computer aided sequence analysis and annotations have made it possible to use EST datasets to study activities of many genes simultaneously and to understand their interaction.

Metabolism of seeds is tailored toward the accumulation of economically valuable storage compounds such as oil and the EST collection therefore provided an opportunity to simultaneously study the transcription of various genes involved in biosynthesis and accumulation of plant lipids. A cDNA library was constructed from developing Arabidopsis seeds (between 5 and 13 days old) actively accumulating storage compounds. After subtraction of the library using highly abundant mRNAs (such as seed storage proteins and oleosins) altogether 10,500 clones were sequenced. These ESTs were annotated and about 40% did not have a match in other databases and thus at the time represented novel genes expressed possibly only in seeds.

Recently, the ESTs have been mapped to the Arabidopsis genome, and they represent about 3500 genes. "Data mining" of this EST data provided insights into the import of photosynthates into developing embryos, its conversion into seed oil, and the regulation of this pathway which could be keys to understand what determines oil content of a seed.

In general the relative abundance of the cDNAs encoding different enzymes of fatty acid synthesis was found to be similar in the seed and non-seed EST sets, suggesting that seeds do not alter to a substantial degree the relative expression of genes encoding pathway components to accomplish the increased flux through the pathway in seeds. Rather, the entire pathway is apparently up-regulated, as suggested by the overall higher relative abundance of ESTs noted for fatty acid synthesis in the seed compared with the non-seed sets. The data suggest that at least nine genes encoding enzymes or subunits involved in this pathway are co-ordinately regulated. This fact would not have been uncovered in studies involving single genes.

ESTs encoding cytosolic glycolytic enzymes were more abundant than ESTs of their plastid counterparts. Transcripts for transporters of phosphoenolpyruvate (PEP) but not for pyruvate could be identified. Together these findings led to a hypothesis that a major route of carbon in developing Arabidopsis seeds involves cytosolic glycolysis from sucrose to PEP, and then PEP is imported to plastids via its transporter and finally converted to pyruvate and acetyl-CoA. This data set also identified >500 known and putative regulatory factors (transcription factors, protein kinases + phosphatases) which represent new, potential candidates for seed-specific metabolic regulators

ESTs in Oilseeds

Soybean and sesame are two important oilseeds for which large EST databases are publicly available. Soybean (Glycine max) is the world's largest source of vegetable oil and a large public genomics initiative has been developed for soybean. As of November 2003 this database contains 344,524 EST sequences originating from over 80 different libraries. This includes ESTs from various sources such as leaves, flowers, shoots, seeds, and somatic embryos. Together they constitute a rich resource for this important agronomic crop.

Sesame (Sesamum indicum) is a highly-valued oilseed crop of Asia that produces seeds with 50% of dry weight oil, and which is commonly used in oriental cooking. Sesame oil also contains lignans that act as antioxidants, stabilising the oil. In order to identify genes involved in storage product accumulation, as well as lignan synthesis, a total of 3328 ESTs from 5-25 days old immature sesame seeds were sequenced. To search for novel or seed-specific sequences, the ESTs were analyzed by BLASTX to the non-redundant protein database of GenBank, and also compared against the Arabidopsis proteome.

The project identified several putative genes or specific isoforms of fatty acid synthesis gene families that may be characteristic to sesame seed. Differences in the expression levels of different isoforms encoding seed storage proteins and oleosins were also observed in comparative analysis of ESTs between sesame and Arabidopsis developing seeds. It was further noted that there is a relatively high expression of cytosolic malate dehydrogenase and low expression of plastidic pyruvate kinase in sesame. These observations suggest that a major pathway of carbon flux for plastid fatty acid synthesis in sesame may involve synthesis of malate in cytosol and its import to plastids, as opposed to PEP import as suggested for Arabidopsis.

ESTs clearly can serve as an important tool for data mining and gene annotations. Comparison of ESTs across different tissues, developmental stages; validated with appropriate statistical analysis could provide insights into regulation of gene(s) and their interactions. However, it is important to recognise various limitations to conclusions based on the EST approach. For example, it is hard to identify genes with low expression levels, such as many transcription factors. Furthermore, even cDNAs of a central fatty acid synthesis enzyme enoyl-ACP reductase were not found in the Arabidopsis EST collection, although the mRNA must be present in seed tissue.

Moreover this approach does not take into account other aspects of regulations such as post-translational modifications, differences in enzyme activities etc. In summary, although EST abundance is an imperfect estimate of gene expression they can lead to useful preliminary insights and hypotheses, which can be further tested by other experiments.

DNA Microarrays for Gene Expression Analysis

The technology of DNA microarrays allows for the first time the simultaneous measurement of thousands of transcripts, and direct comparison of RNA populations from different sources. This technology has revolutionised the ability of researchers to think broadly about gene expression research. Recently, oligonucleotide based arrays have been developed with the complete genome of Arabidopsis represented. Expression profiling of seeds active in fatty acid synthesis has attracted less attention, but a few attempts are follows.

(a) Arabidopsis Developing Seed Microarrays

The first plant fatty acid-related transcriptional profiling project was derived from the MSU Arabidopsis developing seed EST project described above. For the first generation arrays about 3000 DNA elements were chosen for the construction of cDNA based microarrays. These microarrays were initially used for tissue profiling in Arabidopsis to find genes that are expressed highly in seeds by comparing RNA isolated from developing seeds, leaves and roots.

The microarray experiments demonstrated that most seed-expressed transcripts, including lipid biosynthesis related genes, were also expressed in other tissues and so could not be classified seed-specific. Of the >100 lipid genes studied, only 10 were found to be expressed at least 10 fold more in seeds. These genes did not include actual biosynthetic genes, but were more related to FA modification or (such as fatty acid elongase or oleosins). However, a subset of >250 highly seed-specific transcripts were identified, and provided candidates for further studies.

A second generation of Arabidopsis seed cDNA microarrays contained ~6000 DNA elements, which were later mapped to about 3500 Arabidopsis genes. These microarrays were used in time-course experiments to study the temporal patterns of gene expression during Arabidopsis seed filling, as well as comparing gene expression between wild type seeds and seeds from a mutant with a defect in seed storage oil accumulation. Some major observations from this study are listed below:

— Genes associated with different storage products had distinct expression patterns. For example, transcripts for many core fatty acid synthesizing enzymes had a bell-shaped expression pattern, and began to be down-regulated during the active oil synthesis. Examples of this group include ACCase subunits, KAS I and FAD2 (oleate desaturase).

— In contrast, transcripts for some fatty acid modifying enzymes, such as linoleate desaturase (FAD3), elongase (FAE1), as well as for oleosins and storage proteins were induced later and remained high throughout the time course.

— Transcripts for starch metabolism enzymes such as amylases were highest in young seeds, which also transiently accumulate starch.

— Proteins involved in carbon delivery through cytosolic glycolysis via PEP to plastid fatty acid synthesis were co-ordinately expressed. The expression of many proteins (but not all) involved in sucrose import, metabolism and cytosolic glycolysis increased towards seed filling or followed the profile of FAS enzymes. In addition, the expression of PEP transporter followed the profile of the core FAS enzymes, as did the plastid pyruvate dehydrogenase, further strengthening the idea that cytosolic carbon delivery for FAS and fatty acid synthesis enzymes are co-ordinately expressed.

— Several lipid- and carbohydrate metabolism related genes were down-regulated in the wri1- mutant with reduced seed lipid content. Most of these genes belonged to the group with a bell-shaped expression pattern during seed development, and included BCCP, KAS I, FAD2, and chloroplast pyruvate dehydrogenase E1a subunit. In contrast, for genes such as oleosins and fatty acid elongase that are involved in lipid synthesis but did not belong to the bell-shaped expression group were not altered in wri1.

The inserts show changes in relative levels of expression of transcripts between 5 and 13 days after flowering. Microarray expression patterns for enzymes in the cytosol are shown with red lines, reactions in the plastid with green lines and membrane transporters in blue. Grey arrows indicate reactions or transporters whose expression decreases during development and are believed to carry less flux than the major pathway represented by brown arrows. The shift during development from starch biosynthesis and cytosolic metabolism of PEP toward oil biosynthesis and plastid uptake and metabolism of PEP is reflected in the coordinated expression patterns of enzymes and transporters.

(b) Brassica Napus and Maize Embryo FAS Gene Analysis

Recently O'Hara et al. used RNA blots and DNA filter arrays together with EST analysis to study the expression of a group of fatty acid synthesis genes in developing Brassica napus embryos. They found that mRNAs for the biotin carboxylase subunit of plastid ACCase, as well as 3-ketoacyl-ACP reductase, enoyl-ACP reductase, and thioesterase all accumulated in a coordinate manner. Their expression profile peaked around 3 and 4 weeks after flowering, and dramatically decreased afterwards. These expression patterns were similar to the reported mRNA and protein levels of enoyl-ACP reductase and stearoyl-ACP desaturase in Brassica seeds and also to many FAS enzymes in Arabidopsis seeds.

The fact that increased demand for fatty acid synthesis during embryogenesis was met with the up-regulation of a group of synthesis enzymes. In addition, the transcripts were present in constant molar ratio, such that at all times ACP mRNAs were most abundant, and those of the FAS complex were similar to each other's. The results agree with earlier EST comparisons and a study of developing Arabidopsis siliques, which showed that the accumulation of mRNAs encoding the subunits of the heteromeric ACCase subunits is temporally correlated.

Expression of FAS related genes has also been studied from developing maize embryos. Transcript profiles were quantitated using glass based cDNA arrays, nylon membrane arrays, RNA blots, and EST abundance analysis. Similar to previous studies, several FAS genes (ACCAse, KAS III, ACPs, desaturase) exhibited a coordinated expression pattern such that their expression peaked several days before the maximal fatty acid accumulation would be expected. Several oleosin isoforms had a different profile such that their expression was induced at a later stage.

ESTs numbers from libraries constructed at different stages of seed development correlated well with microarray results. The mean correlation coefficient between electronic northern and nylon array expression profiles was 0.86, and between electronic northern and glass array expression profiles was 0.84. This outcome provided an independent confirmation that ESTs and microarray signals both reflect transcript abundance. Together these studies on Arabidopsis, Brassica napus and maize all demonstrate that the up-regulation of several fatty acid synthesising enzymes during seed development seems to be under similar and coordinated control.

Metabolic Engineering of Oilseed

Plant lipid metabolism has proven particularly amenable to modifications both by traditional breeding approaches and more recently by transgenic manipulations. Many technical successes have been achieved, including high-laurate canola; high-stearate canola; reduced saturates and production of gamma-linolenic. Soybeans, canola, sunflower, and cotton have all been successfully engineered for improved fatty acid composition. The production of high-oleic soybeans is described in more detail below because of its potential nutritional and industrial impact.

Although there have been several impressive technical engineering successes, appearance of transgenic oils on the market has been slowed, in part because of concerns regarding antibiotic resistance markers used to produce the original transgenic crops. High-oleic soybean oil: a genetically engineered product with benefits for both food and non-food uses. Most soybean varieties produce an oil rich in polyunsaturated fatty acids - about 50% linoleic acid (18:2) and 10% linolenic acid (18:3). These fatty acids, particularly 18:3, make the oil unstable and easily oxidised. When heated, the oil develops objectionable flavors and odors.

Thus unprocessed soybean oil is unsuitable for many applications, and therefore, for most edible uses it is chemically hydrogenated. This process adds to the cost of the oil and also introduces side reactions such as conversion of double bonds from the cis to trans configuration creating trans-fatty acids. The biosynthesis of polyunsaturated fatty acids in plants is catalysed by a series of enzymes with the first step carried out by an enzyme that converts oleic acid (18:1) to linoleic acid (18:2).

In 1994, the gene (FAD2) for this enzyme was isolated in Arabidopsis by screening mutants generated by T-DNA insertions. Shortly afterward, molecular biologists at DuPont succeeded in isolating and suppressing the expression of the gene in soybean. This strategy led to a major decrease of the 18:1 to 18:2 conversion step and almost completely eliminated polyunsaturated fatty acids in the soybean oil.

The new transgenic soybean oil has 85% oleic acid, one of the highest oleic acid contents found in nature. The absence of polyunsaturated fatty acids eliminates the need for hydrogenation to stabilise the oil. Furthermore, an unanticipated benefit of the oleic increase was that the saturated fatty acid content of the oil fell from approximately 15% to less than 8%. The new soybean oil has a composition similar to olive and other high-oleic oils, which are considered to provide health benefits, compared to other plant and animal oils.

The fatty acid trait was stable in field trials, and the oil yield of the crop was identical to the control lines. Thus neither the transformation process nor the major change in fatty acid composition was detrimental to the high yield of the soybean line. This example is also instructive because it demonstrates how quickly some discoveries can be translated into new crops. With the resources of a major corporation, genetic engineers only needed less than five years from gene isolation to a field-tested transgenic soybean crop ready for commercialisation of a new product.

(c) Health Benefits of High-Oleic Soybean Oil.

The current medical understanding indicates a strong impact of dietary fatty acids on cardiovascular disease and human health. Consequently, there is much interest in tailor-producing healthier vegetable oils and such products may help to balance consumer opposition to "GMO" foods. As mentioned above, health concerns regarding vegetable oil-derived foods include the presence of saturated (particularly palmitic) and trans-unsaturated fatty acids.

Industrial hydrogenation increases saturated fatty acid content and also results in production of trans-isomers of unsaturated fatty acids that are normally not found in vegetable oils and have been associated with coronary heart disease. For many food applications, vegetable oils with a reduced amount of trans-unsaturated and saturated fatty acids are desirable to improve human health. The transgenic high-oleic soybean oil composition provides these benefits in a crop which contributes the major source of fatty acids in American diets. One added consumer benefit to wide future use of the engineered high-oleic oils may be reduction in the pathologies associated with high omega-6 fatty acid consumption, as described above.

Omega-3 Fatty Acid Production in Oil Crops

Due to its prominent role in photosynthetic membranes, the 18 carbon, omega-3 fatty acid, alpha-linolenic is likely the most abundant fatty acid produced in nature. This fatty acid is also found in some oilseeds, although at low concentrations. One notable exception is the oil of linseed (flax) which contains over 50% 18:3. As mentioned above, linolenic acid is easily susceptible to oxidation which initially led to the major use of linseed oil as a "drying oil" in paints, varnishes and other products where the oxidation eventually leads to useful polymerisation. However, food uses of high 18:3 linseed were greatly limited by this tendency to oxidise which also led to negative flavors.

Although oils high in omega-3 polyunsaturated fatty acids are highly susceptible to oxidation and development of flavour and odor problems, the recognition of the health benefits of omega-3 rich fish oils has changed the perspective of some segments of the food industry from emphasizing their elimination toward the development of specialty food oils which are rich in omega-3 PUFA. Such oils would be expected to be marketed as nutritional supplements or as specialty ingredients with a premium price that would reflect the storage/stability issues as well as identity preservation costs and consumer demand.

The beneficial impact of fish oils seems to be most related to their content of the long-chain omega-3 fatty acids docosahexaenoic acid (DHA) and eicosenoic acid (EPA). These fatty acids (or their precursors) are derived from the omega-3 fatty acids of photosynthetic membranes present in algae consumed by fish. Although the 18:3 found in plant oils appears to have some health benefits similar to those of DHA/EPA rich fish oils, the 18:3 presumably must be elongated and further desaturated to influence the eiscosanoid biosynthetic pathways.

Therefore, DHA and EPA are considered more desirable components than 18:3 to promote health benefits. Several groups around the world use genomics to discover genes in marine algae, mosses and fungi that could be used to produce arachidonic acid, DHA or EPA in oilseed crops and a number of research and development efforts are underway to introduce these pathways into plant seeds.

Increasing seed Oil Content

For both edible and industrial uses, an increase in seed oil content is desirable and has been a major goal of oilseed engineering. However, to be economically useful, such a change must not come at the expense of overall seed yield or at the loss of other high-value components. For example, soybean is the largest source of vegetable oil comprising 30% of the world market and now constitutes over 80% of all dietary vegetable oils in the U.S. Although termed an oilseed, soybean contains only 18-22% oil on a seed dry weight basis and is grown principally as a high-protein meal for animal feeds.

Thus, increasing oil in soybean will in most cases not be useful if it comes at the expense of high-value soy protein which drives the crop's economics. By comparison, most other oilseed crops (except cotton) are grown primarily for their oil and produce seeds with 40 to 60% oil. The wide range of seed oil percentage observed in nature suggests that this pathway might be amenable to metabolic engineering, particularly in 'low-oil' oilseeds, provided the key mechanisms which control oil content are identified.

The committed step for de novo FAS is the production of malonyl-CoA catalysed by acetyl-CoA carboxylase. Malonyl-CoA production is a control point for this pathway based upon analysis of acyl-CoA and acyl-ACP pool sizes. Since malonyl-CoA levels in plastids are very low (less than 10%) in comparison to acetyl-CoA, it seemed logical that up-regulation of ACCase activity might increase fatty acid flux. This has been clearly shown to be the case in E. coli. Evidence that increased malonyl-CoA pools could increase fatty acid production was obtained by targeting a homomeric ACCase to rapeseed plastids.

Under the control of a seed-specific promoter this chimeric protein resulted in higher ACCase activities and increased oil yield by 3-5% on a seed dry weight basis. However, the small increase pointed towards additional control points for FAS. Could fatty acid synthase activities also be limiting FAS? Several labs have partly addressed this question by overexpressing enzymes downstream of malonyl-CoA production.

Indeed, overexpression of some activities actually decreased FAS and fatty acid content as observed with the overexpression of a condensing enzyme. Rather than an increase in fatty acid content a 5-10% decrease was observed when a spinach KAS III was expressed in tobacco. Although not all fatty acid synthase enzymes have been overexpressed to determine the effect on FAS, substantial increases in flux will likely require up-regulation of multiple activities.

5

Biopharmaceutical Products from Plants

Plant biotechnology was developed to improve agricultural products, but is now being used to manufacture heterologous proteins for application in industry and human health. The number of proteins successfully expressed in plants is large and is expected to grow rapidly in the future. The types of proteins targeted for this technology include antibodies, food and feed additives, products for human and animal health and industrial enzymes. The use of transgenic plants for large-scale production of proteins (molecular farming) has been shown to be commercially feasible and can be economically advantageous.

Recombinant Protein Production Systems

Many production systems have been used to generate proteins for industrial and pharmaceutical applications. One option is to isolate the protein from its natural source, since proteins thus acquired are usually in their native form and, therefore, functional. However, this approach can be problematic where the source material is expensive, in limited supply or where access is restricted by geography, climate or social issues. If the target protein is a small peptide, chemical synthesis may be a viable alternative to extraction. However, this option is limited to proteins shorter than approximately 50 amino acid residues due both to the cost of synthesis and the difficulty in maintaining structural accuracy in longer molecules.

Foreign proteins have been produced in animal cell cultures and transgenic animals, methods that are very expensive and time intensive, particularly in the scale-up of cultures or herds large enough even for high-value pharmaceutical products (current use), making them highly impractical for industrial enzyme production. Bacteria and fungi are relatively simple fermentative systems but require a large initial investment for capital equipment, and scaling the production systems involves further capital investment.

Choice of Production System

Plants represent an attractive alternative to other expression systems because large amounts of biomass can be produced easily and inexpensively. Transgenic plants require the lowest capital investment (mainly for dedicated harvesting equipment and storage) of all production systems, and they are easily scalable simply by increasing production acreage. The cost of producing crude recombinant protein in plants could be three orders of magnitude lower than that of the mammalian cell system, and 10-fold less than microbial fermentation.

Several additional advantages to plant-based products can be noted, particularly when used for direct delivery. Direct delivery refers to applications where the plant material can be used directly in food, feed or as an industrial feedstock without purification. In some cases, the plant material itself may have value in the application independently of the recombinant proteins contained in it. Furthermore, plant systems allow for proteins free from human pathogens.

Technology Options

The choice of plant to be utilised depends on a number of factors including its cultivation, transformability, growing cost, production and processing of the target tissue, existence of wild relatives, and degree of out-crossing. Current systems include corn, soybean, canola, alfalfa, Lemna, tobacco, tomatoes, potatoes and safflower. These crops are examples of wild plants, e.g., Lemna, domesticated non-food crops, e.g., tobacco and alfalfa, and food crops-rice, corn, soybeans, tomatoes, potatoes, canola, and safflower.

The type of tissue that is utilised for protein accumulation is often chosen because of the type of plant used, or vice versa. For example, Lemna, alfalfa and tobacco are leafy crops, and the seed or fruit would not be appropriate as production vehicles. Potatoes are a root crop and thus the tubers are used. For fleshy fruits, tomatoes can be used for production. However, grain crops such as corn, rice, soybeans, canola and safflower have the most stable production vehicle, the seed. While multiple factors affecting protein accumulation and crop choice are interrelated, only generating the transgenic material and protein expression will be discussed.

Transgenic Material

Recovery of stably transformed plants exhibiting traits of interest requires the combination of several technologies: 1) appropriate selectable marker genes and selection conditions, 2) an efficient culture system that allows recovery of plants from target tissues, 3) a DNA delivery system, and 4) joining of these technologies so that transformed, fertile adult plants can be recovered. Variety is the driver for these technologies because of the plethora of plant species available as targets for plant production of foreign proteins. For maximum utility and efficiency, however, DNA delivery systems should be simple, efficient and preferably inexpensive.

Molecular farming requires that DNA for encoding the protein of choice be introduced at will into the plant of choice. To this end, multiple methods of foreign DNA transfer into plant nuclei have been developed over the last 20 years. These methods include Agrobacterium-mediated transformation and several direct gene transfer methods, e.g., microprojectile bombardment, electroporation, silicon carbide fibers, electrophoresis and microinjection.

Microprojectile bombardment has also been used to transform chloroplasts of solanaceous plants. Of these numerous methods, the Agrobacterium-mediated nuclear transformation and microprojectile bombardment-mediated nuclear and chloroplast transformation methods are most often used today. Viral vectors are being used for commercial development of transiently expressed therapeutic proteins in tobacco.

The operative word here is variety. Crop plants span many species in many genera, families and classes, and no single method is appropriate for gene transfer. Selection of transformed tissues requires inclusion of genes that allow identification of the transformed cells. Selectable marker genes come in a variety of types with quite varied substrates. Again, the operative word is variety, because a large array of methods and markers are critical to the success of plant biotechnology.

Selectable markers can provide selection pressure on plant tissues resulting in death of non-transformed cells, or through the starvation of unwanted cells because selective growth of transformed cells is supported. Selectable markers have recently been deemed as undesirable traits because of their perceived danger to public health. Their perceived danger is because of the potential for allergenicity of the protein or the potential of resistance gene transfer to gut microorganisms (antibiotic resistance).

Maintenance of selectable marker genes in the plant after establishment of the desired trait is assumed to be unnecessary. The assumption is that these genes have no utility after transformed plant recovery. However, herbicide resistance genes have utility because they confer downstream advantages for selection of transformed plants in the field.

Herbicide resistance gene products allow growth and recovery of plants prior to establishment of homozygous lines so that early field performance of the transgenic plants can be assessed. Moreover, the pflp gene, a selective trait, confers resistance to plant pathogens, also giving it value beyond selection. Thus, a discriminating assessment of marker gene value versus risk should be undertaken before wholesale removal of the trait is endorsed.

Gene Expression and Protein Accumulation

Gene expression with its resulting protein accumulation is the single most important factor for expressing most recombinant proteins. The economics of the product as well as regulatory issues will be dictated by gene expression patterns and amounts. Expression of the gene depends on many factors including transcription, translation, targeting and the ability of the plant to accumulate the protein.

In the case of foreign protein production of pharmaceuticals and industrial enzymes, it is desirable to sequester the protein as much as possible into specific target tissues. This is accomplished with the use of tissue specific promoters that are active only in limited tissue types. An example is the globulin-1 promoter from maize that is primarily limited to embryo specific expression.

In addition to promoter specificity, the subcellular compartments that are noteworthy within that tissue present the array of potential subcellular locations that are likely to maximise that promoter's work. An example of the effect of alternative targeting on protein accumulation is shown in Table 1. Clearly, cell wall targeting effected the highest accumulation of protein 1, whereas the vacuole was the best subcellular location for accumulation of protein 2. The cellular machinery responsible for targeting proteins is under intense study and has been elegantly reviewed.

The rough endoplasmic reticulum (RER) contributes protein to the several compartments that receive their member proteins from the membrane/secretion pathway, e.g., the vacuole, plasma membrane, the endoplasmic reticulum (ER), the Golgi and the cell wall. Secretion to the exterior of the plasma membrane is the default pathway, and a signal sequence that begins this process is generally necessary and sufficient to have a protein arrive on the cell surface. Other destinations in this pathway require additional information.

The plastids, mitochondria, nucleus and cytoplasm receive their proteins from cytoplasmically translated messages, and the proteins have specific transit peptide sequences that allow their import into each organelle. The plastids are an interesting group of organelles that are derived from a single pre-organelle, the proplastid, and include chloroplasts, chromoplasts, amyloplasts and elaioplasts. The import process for chloroplast proteins encoded by nuclear genes has been reviewed recently.

The underlying principle is that a transit peptide is necessary for recognition by a receptor on the surface of the plastid envelope. Some features of this transit peptide are common to all plastids. However, each member of the plastid group most likely has features of its import apparatus that limit uptake of proteins to those specifically required for the unique functions of the various plastids.Transit peptides and signal sequences have been utilised by all players in the field of molecular farming. In most cases, these signals are a part of the strategy to achieve maximal accumulation of the target protein, which is the measured parameter.

Table 1. Effect of promoter and targeting signal on protein accumulation in maize seed

Protein location	Protein example	Targeting sequence	Highest T1 seed
Embryo [a]Cell wall	1	BAASS	0.8% TSP
Constitutive[a] Cell wall	1	BAASS	0.19% TSP
Constitutive[a] ER	1	BAASSKDEL	0.12% TSP
Constitutive[a] Cytoplasm	1	None	0.07% TSP
Constitutive[b] Vacuole	2	SSVacuole	10% TSP
Constitutive[b] Cell wall	2	BAASS	2% TSP
Constitutive[b] ER	2	BAASSKDEL	0.08% TSP
Constitutive[b] Cytoplasm	2	None	0.0008% TSP

[a] Data taken from Hood et al., 2003

[b] Data taken from Streatfield et al., 2003

In very few cases has subcellular in situ localisation of the target protein been performed. In one case, the bacterial endotoxin from E. coli (Lt B) was found in an unpredicted compartment. In order for proteins to accumulate, the protein must be stable to the particular environment in which it is found. Carbohydrate, protein and lipid content, as well as pH and salt may influence the stability of the protein. Since these will differ in different plants, tissues and subcellular locations, protein accumulation will also vary considerably. While this must be explored empirically today there is hope that the future may bring predictability to the fate of proteins based on their specific characteristics in plants and tissues to achieve the highest overall accumulation.

Production Practices

Production practices for recombinant proteins refers to the growing, harvesting, transport, storage and tissue processing of the crop, as well as the extraction and purification of the protein (Table 2). With thousands of species to select from and the wide variety of products that are possible, it is highly unlikely that any one system will work best for each of these steps. Moreover, different product categories will require different amounts of purification and different formulations.

Each plant production system has distinct characteristics that may prove advantageous depending on the product. The key features of the best production system include a potential for low cost of goods, maintenance of protein integrity, flexibility with regard to time and temperature for harvest and maintenance of product as well as environmental safety. In selecting a crop for production one must decide if working with a domesticated species would be preferred over use of a wild species (Figure 1).

Native plants only grow well in specific environments because of light, temperature and soil conditions. In contrast, the major commercial crops have been adapted for a wide range of conditions, which extends their geographic boundaries. In many cases only one crop per year can be grown in a geographic area. However, additional production of this same crop can be done in other geographic locations at different times of the year, extending the seasonal growing of the crop to year-round production.

Table 2. Steps in Production with Key Considerations

Production step	Consideration	Applicable products
Growing	Requirements Geographic Limitations Seasonal Limitations Recombinant Protein YieldConfinement	
Harvesting	Mechanical Issues Time Sensitivity	
Transportation	Temperature sensitivity Cost	
Storage	Temperature sensitivity Protein stability in tissue	First critical
Tissue Processing	Stability in tissue Potential for Enrichment	step for specific product type[a]
Extraction	Protein stability Biomass Quantity	Not necessary for direct delivery
Purification	Interfering agents CGMP	Primarily for pharmaceuticals

[a] Vaccines, industrial enzymes, purified pharmaceuticals

Wild species have an advantage in that they will be unlikely to be mistaken for food crops and therefore unlikely to be inadvertently mixed with the food supply. Unfortunately, they would be more likely than food crops to outcross with native species. One consideration for foreign protein production is whether to use an open-pollinated or a self-pollinating plant, an issue for either domesticated or wild species. For most domesticated crops, self-pollination usually means that the seed planted by the grower can be saved every year and replanted the following season. This is in contrast to open-pollinated crops that are produced as hybrids. For these, subsequent production is not from plants in the production fields but from parent seed stocks.

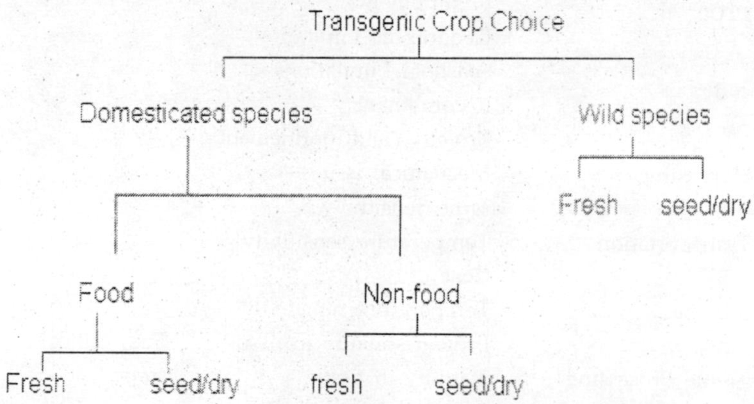

Figure 1. Characteristics of plant systems

With the exception of some specialty crops, most of the crops today are harvested mechanically. After the plant tissue is harvested and transported to its designated location, it must be stored for some amount of time before it is processed. Leaves and fruits contain water, making storage problematic. In contrast, if a plant storage organ is used, e.g., seeds or tubers, the plant part is in a dormant state with little metabolic activity, and storage is easy.

Tissue processing after the crop is harvested is necessary whether for direct delivery or for highly purified products. One of the most critical aspects is the amount of total protein present in the harvested tissue.

The amount of protein as a percentage of total biomass can range from less than 1% to over 40% depending on the plant and tissue source. This feature is critical because in addition to obtaining relatively high percentages of total soluble protein for ease in purification, the overall cost of tissue processing and extraction is directly related to the amount of necessary biomass to obtain the required product amount.

Tissue can also be processed and separated into fractions enriched with recombinant proteins by mechanical means. As an example, it is possible to generate a germ or endosperm fraction from grains by using standard procedures common in the industry today. These fractions can have much higher protein content than the whole grain because the expression technology used may sequester the protein in a specific sink. Furthermore, the remaining part of the grain can be used for other industrial applications.

Extraction is relatively easy in most cases by simply adding an aqueous buffer to ground tissue. The pH and salt content can be optimised on a case by case basis to reduce extraction of endogenous proteins and preferentially solubilise the recombinant protein. Fresh tissue will most likely require refrigeration to prevent protein degradation. Seeds, however, usually have the advantage of endogenous protease inhibitors, which allows greater flexibility in extraction times and temperatures.

Protein can be purified from the aqueous extracts in a manner similar to extracts from other production systems. The advantages and disadvantages for a variety of plant types suggests the possibility that one could modify an existing crop specifically for recombinant protein production. As an example, a cultivated crop could be altered to have a higher protein content that could translate into higher amounts of recombinant protein.

For industrial feedstocks, selecting a major crop that already is used in industrial applications would have benefit. For orally delivered therapeutic proteins or vaccines, a food crop that has GRAS (Generally Recognised as Safe) status would be best. For protein stability and ease of transport a grain would be a good choice.

Human Health Products

High purity human health products made in plants have the distinct advantage of being free from pathogens that are inherent in animal production systems and from the contaminants of microbial systems. The actual purification procedures and costs after extraction of the plant material are similar to these systems. Trypsin and aprotinin manufactured in transgenic maize are examples of high purity products from a plant system. Each of these products had special requirements to achieve high expression. Because trypsin is an aggressive protease, the zymogen form was expressed in the seed and these parameters were shown to be essential for success. Aprotinin expressed from a seed-preferred promoter with the protein accumulating in the cell wall space generated extremely high-expressing lines. Each protein can be produced from a small acreage to fill the required market amounts.

Protection against infection by transmissible gastroenteritis virus (TGEV) has been demonstrated following oral delivery of a subunit vaccine produced in transgenic maize. This was the first demonstration of disease protection following this oral delivery methodology. Again, the major advantage is freedom from human and animal pathogens in the production system. However, the other advantages include low overall cost of production and utilisation of a feedstuff directly from a crop.

Industrial enzymes are ideal for use as catalysts in industrial processes because they can function at ambient temperatures in aqueous solutions at neutral or physiologic pH levels. Furthermore, enzymes have a high specificity, low toxicity, and catalyse reactions relatively quickly at low concentrations. However, temperature and pH extremes can limit enzymatic activity, and because these extremes are present in large-scale industrial processes, significant changes are required when moving from chemical to enzymatic catalysts.

Nevertheless, the large market opportunity for industrial enzymes and the environmental benefits they provide offer an incentive to invest in the development of efficient production systems and to make changes in industrial processes to accommodate enzymatic catalysts. Cost models indicate that industrial enzymes can be produced cost-effectively at 0.1% of grain dry weight. With today's technology it is possible to obtain 1% of grain dry weight as a single recombinant protein (Figure 2). Because plant biotechnology is young, improvements will be achieved over the next several years that will enable substantial increases in recombinant protein expression.

Laccase and manganese peroxidase represent two examples of industrial enzymes produced in maize. These two oxidation/reduction enzymes have many potential applications in industries including wood products and textiles because of their activities on lignin. Laccase is actually more active as a ground corn fraction than in an extract, whereas manganese peroxidase is easily extracted and concentrated. Application of these enzymes in industry has been hampered by the ability to produce

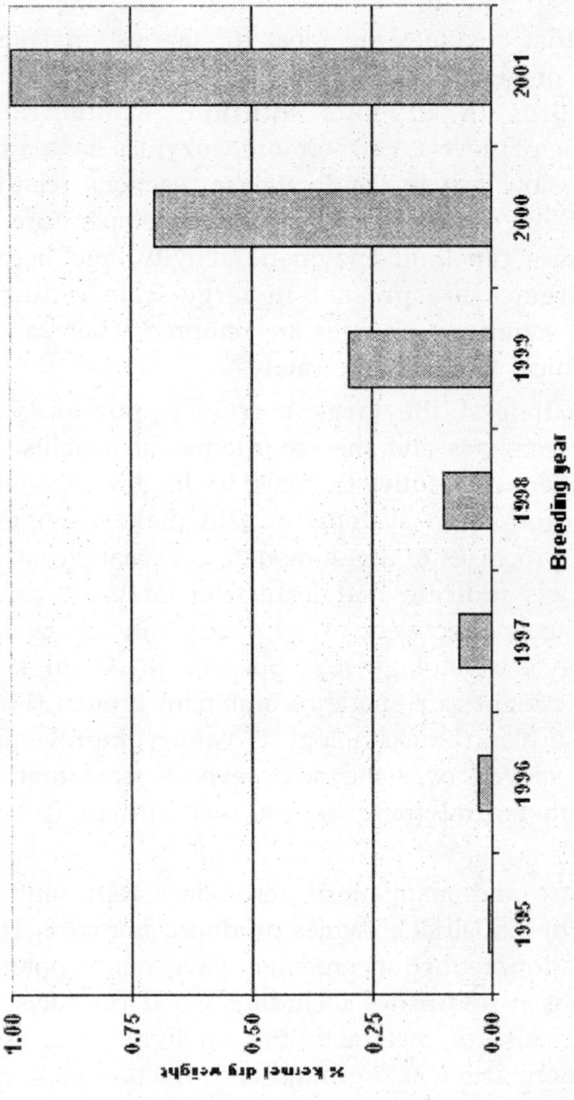

Figure 2. Increase in avidin expression in maize kernels by breeding and selection

them economically in large quantities. Their high expression in transgenic maize should enable this process.

Other issues

Production is composed of the mechanics of product development, manufacturing and formulation. However, in order to bring a product to market, several other considerations are necessary.

Legal issues. A consideration of legal issues surrounding any new industry is important. In the case of plant molecular farming, legal considerations include such things as patent protection of intellectual property and licensing of patented technology to provide freedom to operate. Consideration should also be given to a strategy for invention versus licensing. In many cases, it may be more expensive to invent new technology than to license existing technology. The issues are complex, and expert legal advice is key to success in this field.

Regulatory issues. Regulatory issues are becoming more complex. Regulations that are based on scientific principles rather than reaction to public fears are the goal of scientists as well as the regulatory agencies. However, in order for the agencies to accomplish that, the data must be collected. Never have agricultural products undergone such scrutiny, whether food products or crops containing pharmaceuticals or industrial enzymes.

Public acceptance. The public must see a positive benefit from genetically modified crops in order for acceptance to be a reality. Product safety is also at issue here, whether the crop is a foodstuff with herbicide or insect resistance genes, or a crop containing a pharmaceutical or industrial product. The safety of the crop can be assessed quantitatively using models for assessing risk/benefit ratios.

6

Myriad Plant Responses to Herbivores

Plants encounter multiple biotic and abiotic challenges simultaneously, in their natural habitat. Each environmental hazard activates multiple signal transduction pathways to ensure an effective spatial and temporal defense response. Therefore, plants must be able to identify and prioritise each signaling pathway to mount the most efficacious defense strategy to minimise current and future damage and also to preserve vegetative growth and reproductive success. These complex biochemical and physiologic responses often result in a tolerance or protection from further environmental challenges.

To protect themselves from pathogen and herbivore attack, plants use constitutive and induced defenses. These defenses can influence herbivore settling, feeding, oviposition, growth and development, fecundity, and/or fertility. All defenses, whether constantly or transiently expressed, are costly. Defense responses channel carbon and nitrogen resources from vegetative and reproductive growth into protective mechanisms. The plant must attain a balance to ensure survival from immediate and subsequent attacks without sacrificing plant vitality, longevity, or reproduction. The balance of constitutive and induced defense responses appears to supply the plant with the flexibility to achieve these goals.

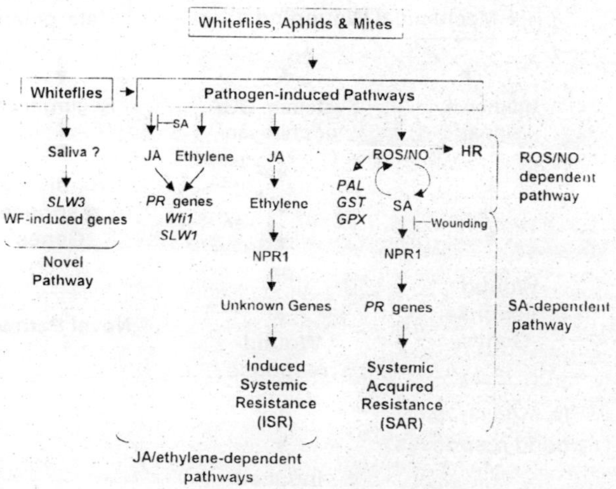

Figure 1. The signal transduction pathways induced by pathogens and herbivores are outlined

Constitutive defenses include the physical barriers that impede pathogen ingression or arthropod access to tissues, that is, cell walls, suberin, callose, and cuticles, as well as the stored allelochemicals that have antixenotic (deters herbivore colonisation of plant) or antibiotic (deters herbivore growth, reproduction, development, or survival) effects. Constitutive defenses, as well as induced defenses, are only effective if the herbivore contacts the defense phytochemical. Because allelochemicals are often species-specific and are expressed in a subset of tissues or cells, the choice of host plant by an herbivore and the mode and site of herbivore feeding determines the defense chemicals that are encountered.

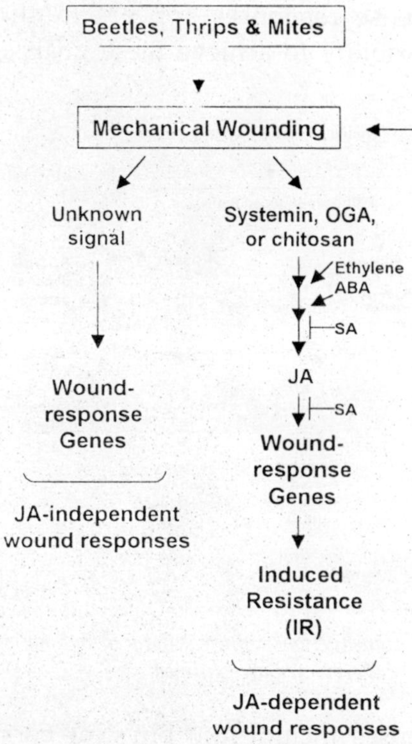

Figure 2. The signal transduction pathways induced by wounding and herbivores in the Solanaceae are illustrated

Plant responses to herbivores that cause less tissue damage such as arthropods that mine or gall, or herbivores that pierce or lacerate cells to feed on intracellular fluids. Herbivores induce several wellcharacterised plant defense- and wound-response pathways, as well as novel pathways to alter plant gene expression (Figures 1, 2). Herbivores produce novel signals (elicitors) to activate plant gene expression and volatile synthesis.

Activation of Pathogen-defense Response Pathways

Insects that use a piercing/sucking mode of feeding have an intimate and long-lasting interaction with plant cells. Using a stylet to pierce cells, these herbivores consume large quantities of fluids as a nutritional source. Feeding sites vary. Most aphids, mealy bugs, leafhoppers, psyllids, and whiteflies stylets must traverse the cuticle, epidermis, and mesophyll to establish feeding sites in veins of the phloem. Other piercing/ sucking insects primarily feed on (1) mesophyll parenchyma (scale insects), (2) both epidermal and mesophyll cells (thrips), or (3) xylem (leafhoppers). The amount of tissue damage caused by piercing/ sucking insects varies tremendously. Some herbivores (like thrips and spider mites) are cell-content feeders; they lacerate cells and consume cellular contents by means of their stylets. Other cell-content feeders such as the pyrrhocorids and lygaeids (truebugs) cause more extensive damage along the path to and at their feeding site.

The stylets of piercing/sucking insects that feed on phloem are in continuous contact with plant cells. Once feeding sites are established, they can be used for hours to weeks. Given the limited tissue damage and prolonged stylet interactions with plant cells, it is not surprising to find that plant responses to phloem-feeding insects are distinct from that of chewing insects and tissue-damaging cell-content feeders. On the basis of the limited number of studies currently available, some piercing/sucking insects induce the defense-signaling pathways most commonly activated by bacterial, fungal, and viral pathogens (Figure 1).

Pathogen Attack and Plant Responses

Signaling mechanisms activated after pathogen attack

have been elegantly dissected using the tools of genetics, molecular biology, and biochemistry. When pathogens and plants with cognate avirulence (avr) and resistance ® genes interact (incompatible interactions), pathogens are rapidly perceived. Plants respond with production of reactive oxygen species (ROS) and nitric oxide (NO), membrane depolarisations, ion fluxes, and activation of signaling cascades that involve the action of kinases and phosphatases. The signaling cascades result in the accumulation of defense-response RNAs and proteins locally and systemically (Figure 1; Table 1).

Table 1. List of Genes and Their Roles in Plant Defense or Wound Responses

Genes	Function in Defense or Wounding
Tomato *Chi3*	*Chi3* encodes an acidic chitinase that hydrolyzes chitin in pathogen cells walls and possibly in insect guts. Acidic chitinases are localised in the apoplast.
Tomato *Chi9*	*Chi9* encodes a basic chitinase that hydrolyses chitin in pathogen cells walls and possibly in insect guts. Basic chitinases are localised in the vacuole.
Tomato *GluAC*	*GluAC* encodes an acidic β-1,3-glucanase that hydrolyzes callose and glucan polymers of pathogen walls. Acidic glucanases are localised to the apoplast.
Tomato *GluB*	*GluB* encodes a basic b-1,3-glucanase that hydrolyses callose and glucan polymers of pathogen walls. Basic glucanases are stored in the vacuole.
Tomato *LapA*	*LapA* encodes the wound-induced leucine aminopeptidase that removes *N*-terminal amino acids from peptides and proteins. *LAP-A* function in wounding is not known. It is only found in a subset of the Solanaceae (Figure 4).
Tomato *LOX* Arabidopsis	LOX encodes the 13-lipoxygenase that synthesises 13-hydroperoxide-

Contd . . .

Contd . . .

LOX1, LOX2	octadecatrienoic acid from linolenic acid (an 18:3 fatty acid), (Figure 4).
Tomato *PAL* Arabidopsis PAL	PAL (phenylalanine ammonia lyase) is important for the biosynthesis of SA, flavanoid phytoalexins, lignin, and other cell wall phenolics.
Tomato *PR-1*	The *PR-1* gene of tomato encodes an acidic PR-1 protein also known as P4 and highly related to P6. Its function is not known. Acidic PR-1 has an apoplastic location. Basic *PR-1* genes also exist and encode proteins with a vacuolar location.
Tomato *P2*	The P2 protein is a basic win-like protein with a chitin-binding domain. Its function is unknown and it has a vacuolar location.
Tomato *PR-4*	*PR-4* encodes an acidic win-like protein with a chitin-binding domain. Its function is unknown and it has an apoplastic location.
Tomato *pin2*, pin1 Arabidopsis PIN	*Pin* genes encode inhibitors of Ser proteases. Pins interfere with insect growth and development by hyper-inducing proteases in the insect gut (Figure 4).
Tomato *PG*	A wound-induced PG (polygalacturonase) hydrolyses pectin in the cell wall to release OGAs, which are potent signals that activate the tomato octadecanoid pathway (Figure 4).
Tomato *Sys*	*Sys* encodes prosystemin, which is a precursor protein that is proteolytically processed to its bioactive peptide systemin (Figure 3). Systemin is only found in Solanaceous plants. Systemin is a potent activator of the octadecanoid pathway (Figure 4).
Tomato *Wfi1*	Wfi1 is the large subunit of the multi-subunit plasma-membrane complex called NADPH oxidase. NADPH oxidase generates the reactive oxygen species superoxide anion. Only the gp91-*phox* subunit has been cloned in plants (Figure 1).

Contd . . .

Contd . . .

Arabidopsis ACO	*ACO* encodes an acyl CoA oxidase-like protein. ACO is involved in b-oxidation of fatty acids. Its substrate and function in the wound response are not known (Figure 5B).
Arabidopsis AOS	AOS (allene oxide synthase) converts linolenic acid (18:3) to a 13-hydroperoxide form. This enzyme commits lipids to the octadecanoid pathway and appears to have an important regulatory in regulation of the wound responses in *Arabidopsis* (Figure 5B).
Arabidopsis CHS	*CHS* encodes chalcone synthase important for the synthesis of lignin to strengthen the cell wall and phenylpropanoid compounds that can have antimicrobial activity (Figure 5B).
Arabidopsis CK	*CK* encodes a choline kinase–like protein. It function in the wound-response is not known (Figure 5B).
Arabidopsis CPR1	CPR1 (constitutive expresser of PR genes) function is not yet known. It is a regulator of multiple defense signaling pathways. *cpr1* mutants express *PR* proteins and SAR constitutively (Figure 5B).
Arabidopsis DFR	*DFR* encodes dihydroflavanol reductase. DFR is important for the synthesis of anthocyanin pigments (Figure 5B).
Arabidopsis GST	*GST* encodes glutathione S-transferase. GST is important in detoxifying many compounds and ROS (Figure 5B).
Arabidopsis GPX	*GPX* encodes glutathione peroxidase. GPX is important in scavenging of ROS (Figure 5B).
Arabidopsis JR1	*JR1* (*JA-regulated 1*) encodes a protein of unknown function (Figure 5B).
Arabidopsis JR3	*JR3* (*JA-regulated 3*) encodes a protein similar to the *ILR1* amidohydrolase. ILR1 releases auxin from conjugated forms. The role of JR3 in wounding is not known (Figure 5B).
Arabidopsis	*NPR1* (nonexpresser of *PR* genes) encodes
NPR1	protein similar to the transcription
IKB.	factor inhibitor
	This is a critical regulator of both SA-induced responses and responses leading to ISR. *npr1*

Contd . . .

Contd . . .

	(also known as *nim1*) mutants cannot induce *PR* gene expression or SAR (Figure 1).
Arabidopsis PDF1.2	*PDF1.2* encodes the small antimicrobial protein defensin. *PDF1.2* is regulated by JA but not by wounding.
Arabidopsis SSI1	The function of SSI1 (suppressor of SA-insensitivity) is not known. The *ssi1* mutant suppresses the *npr1* phenotype. SSI1 is a key regulator of multiple defense signaling pathways.
Arabidopsis TAT	*TAT* encodes a tyrosine aminotransferase that is important for the synthesis of tyrosine. Tyrosine is used for the synthesis of cell wall phenolics, lignins, and flavanoids (Figure 5B).
Arabidopsis Thi2.1	*Thi2.1* encodes thionin, which is a small polypeptide with anti-fungal activity (Figure 5B)
Arabidopsis VSP	*VSP* (vegetative storage protein) encodes a protein that accumulates to high levels in leaves. *VSP* is regulated by JA, but its role in defense is not known (Figure 5B).
Arabidopsis WR3	*WR3 (wound-response 3)* encodes an RNA that accumulates in response to wounding but its function is not known (Figure 5B).
Squash *SLW1*	*SLW1 (silverleaf whitefly-induced 1)* encodes a M20b peptidase. Its function unknown. *SLW1* is preferentially induced by silverleaf whiteflies and is regulated by JA and ethylene (Figure 1).
Squash *SLW3*	*SLW3 (silverleaf whitefly-induced 3)* encodes a β-glucosidase-like protein. Its function unknown. *SLW3* is preferentially expressed in response to silverleaf whiteflies and is regulated by a novel signaling pathway (Figure 1).

Defenseresponse proteins hydrolyse pathogen cell wall polymers, strengthen and modify plant cell walls, turnover proteins, enhance synthesis of secondary metabolites, generate signals to modulate defensesignaling pathways, or have unknown functions in defense. Similar biochemical events are induced in compatible interactions but at a slower pace. Incompatible (fast recognition) and compatible (slow recognition) interactions are also distinguished by the presence or absence of a hypersensitive response (HR), respectively.

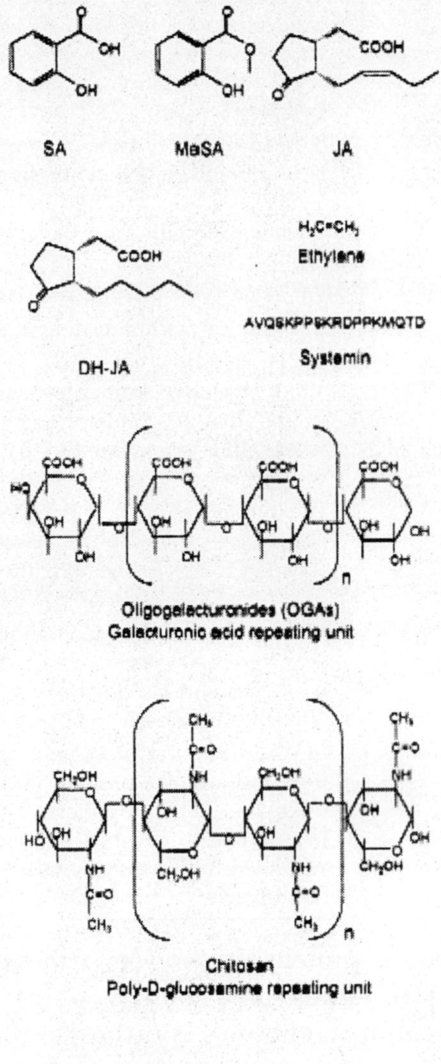

Ctd . . .

Ctd . . .

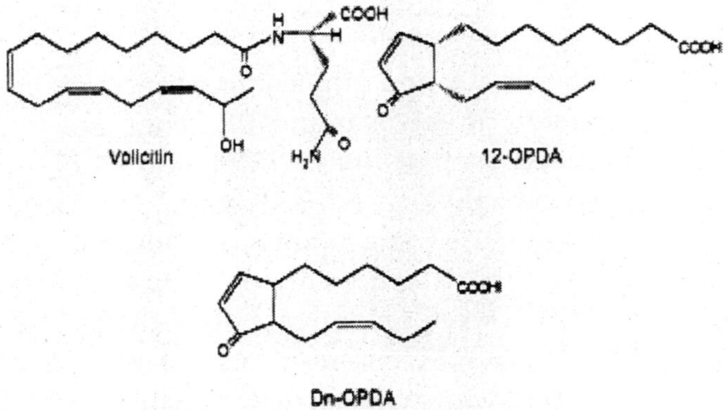

Figure 3. Structures of signaling compounds in herbivore-plant interactions

During incompatible interactions, the HR causes a rapid, localised cell death at the site of infection, which restricts the pathogen to small areas of tissue. Micro-HRs occur systemically and may aid in activation of defense responses at remote locations on the plant. Four signaling pathways are important in responses to pathogens (Figure 1). First, the salicylic acid (SA)-dependent cascade uses SA (Figure 3) and its methyl conjugate (MeSA; Figure 3) to stimulate expression of defense-response genes, including pathogenesis-related protein (PR) genes that encode proteins with an apoplastic localisation (acidic PR genes) (Table 1).

SA also promotes the development of systemic acquired resistance (SAR), which confers a broad-range resistance to pathogens and some insects. The second pathway is dependent on ROS and NO, which increase after pathogen attack. These molecules promote an HR, stimulate SA synthesis, and induce some defense-response genes (Figure 1, Table 1). The remaining two pathways are regulated by jasmonic acid (JA) and ethylene (Figure 1). Both JA (Figure 3) and ethylene (Figure 3) levels increase after pathogen attack.

In Arabidopsis, JA and ethylene act concomitantly to induce expression of defensin and PR genes that encode vacuolar-localised proteins (basic PR genes) (Table 1). In addition, JA and ethylene act sequentially to induce a systemic tolerance to a broad range of pathogens, which is distinct from SAR, and called induced systemic resistance (ISR). Development of ISR is not correlated with the SAor JA/ethylene-induced genes studied to date. Several other uncharacterised signaling pathways may be active after pathogen attack.

The SA- and JA/ethylene-dependent signaling pathways cross talk. Rises in SA are correlated with down-regulation of the JA/ethylene-regulated defense-response genes, as well as JA-regulated wound responses (Figures 1, 2). Coordination of these defense pathways with each other and with wound-response pathways is complex and not understood at this time (Figures 1, 2). These signaling mechanisms appear to converge at several regulatory junctions involving wound-induced (WIPK) and SA-activated (SIPK) MAP kinases. In addition, some gene products that regulate PR gene expression, NPR1 (non-expresser of PR genes), SSI1 (suppressor of SA insensitivity), and CPR6 (constitutive expresser of PR genes), influence multiple signal transduction pathways. Convergence of these signaling networks allows plants to prioritise cues and activate the mechanisms that most effectively combat the current invading pathogen or pest.

Defense-signaling Pathways by Herbivores

Several nonchewing arthropods increase levels of PR proteins and activities (Table 1). Increases in chitinase or b-1,3-glucanase activities after whitefly, aphid, or mite infestations have been observed. Increases in chitinase and β-1,3-glucanase activities are correlated with increases in these proteins. Furthermore, increases in P2, P4, and PR-10-like proteins are detected. As in incompatible plant-pathogen interactions, PR proteins and activities increase more rapidly when herbivores attack resistant plant genotypes.

Feeding by greenhouse whitefly (Trialeurodes vaporariorum) and silverleaf whitefly (Bemisia argentifolii) nymphs activates the SA- and JA/ ethylenedependent pathways of tomato (Figure 1; Table 1) (Puthoff and others unpublished). However, PR gene RNAs do not increase locally or systemically in response to adult whiteflies. Transcripts for PR genes regulated by *JA* and/or ethylene (basic β-1,3-glucanase, basic chitinase, and *PR-1*) accumulate to higher levels than *SA*-regulated gene RNAs (acidic β-1,3-glucanase and acidic chitinase)

Similar to whiteflies, the potato aphid (Macrosiphum euphorbiae) and green peach aphid (Myzus persicae) cause increases in lipoxygenase (LOX) and *PR-1* RNAs in infested tomato leaves. In Arabidopsis, the cabbage aphid (Brevicoryne brassicae) and the cotton aphid (Aphis gossypii) induce both the SA-dependent (*PR-1* and acidic β-1,3-glucanase) and JA/ethylenedependent (defensin and LOX) signaling pathways.

Consistent with limited tissue damage during whitefly feeding, wound-response gene RNAs (leucine aminopeptidase [*LapA*] and proteinase inhibitor [pin2]) (Table 1) do not accumulate in response to adult or immature whitefly feeding (Puthoff and others unpublished). Furthermore, analyses of LapA:GUS transgenic tomato plants show that the *LapA* promoter is not activated in cells along the stylet path (Puthoff and others unpublished). Likewise, phloem-feeding aphids do not increase *pin2* RNA levels. This contrasts to the tomato response to the two-spotted spider mite (Tetranychus urticae) and western flower thrip (Frankliniella occidentalis). Although these arthropods use stylets to feed, they cause substantial cellular damage and induce JA-dependent wound-response genes of tomato (G. Howe, personal communication).

Whitefly Species and Developmental Stages

Studies on the whitefly-induced 1 (*Wfi1*) gene of tomato and silverleaf whitefly-induced 1 (*SLW1*) and *SLW3* genes from squash indicate that some plant genes respond to signals generated by specific insect species and specific stages in insect development (Figure 1). Wfi1 RNAs accumulate locally and systemically in tomato leaves after feeding by whitefly nymphs but not in response to pink potato aphids or whitefly adults.

These data suggest that some of the signals generated by whitefly and aphid feeding, as well as adult and immature insects, are distinct. Wfi1 encodes a membrane-bound subunit of NADPH oxidase (a gp91-phox homologue) (Table 1) (Puthoff and Walling unpublished). Increases in NADPH oxidase activity are correlated with the ROS burst that accompanies pathogen infection and wounding. Like basic PR genes, *Wfi1* RNAs accumulate after JA and ethylene treatments but not after systemin treatments (a potent inducer of wound-response genes) (Figures 1, 2).

Surprisingly, wounding transiently increases levels of Wfi1 RNAs. These data indicate that Wfi1 is primarily regulated by the JA/ ethylene defense-signal transduction pathway after whitefly infestation (Figure 1) and may be activated by a systemin-independent pathway after wounding (Figure 2) (Puthoff and others unpublished). Even closely related insect species, such as the silverleaf whitefly and the sweetpotato whitefly (B. tabaci Type A), produce elicitors that differentially regulate plant gene expression (van de Ven and others 2000). Similar to tomato PR genes and *Wfi1*, *SLW1*, and *SLW3* transcripts increase in response to silverleaf whitefly nymph feeding but not to adult feeding (Table 1).

More notably, *SLW1* and *SLW3* RNAs accumulate systemically after feeding by the silverleaf whitefly nymphs but not after feeding by sweetpotato whitefly nymphs. The preferential expression of SLW1 and SLW3 by the silverleaf whitefly suggests that this insect generates (1) a novel signal, (2) larger amounts of a systemic elicitor, or (3) a more potent signal than is produced by the sweetpotato whitefly. Treatments with defense signals show SLW1 is regulated by both JA and ethylene (Figure 1).

In contrast, *SLW3* RNAs do not accumulate in response to any known defense/wound signals (NO, H_2O_2, NO plus H_2O_2, JA, ethylene, SA, abscisic acid, JA plus ethylene, JA plus SA, or wounding). These data suggest that the silverleaf whitefly activates a novel signaling pathway to induce *SLW3* expression (Figure 1). Interestingly, both SLW1 and SLW3 transcripts accumulate after waterdeficit stress, suggesting that the signals produced by silverleaf whitefly feeding and water-deficit may share some features. This is in contrast to caterpillar feeding in Arabidopsis, where genes induced by water-deficit are expressed in response to wounding but not to caterpillar feeding.

Elicitors in Aphid-, Mite- and Whitefly-plant Interactions

The signals generated by herbivores, which pierce plant cells to remove liquids as a nutrient source, are complex. Some cues, like those that activate PR gene expression, are likely to be shared by many herbivores (whiteflies, aphids, and mites). In contrast, some elicitors appear to regulate species-specific responses such as the changes in levels of *Wfi1* and *SLW* RNAs and the complex volatile blends that regulate plant-herbivore-herbivore enemy interactions. These general or specific signals may derive from physical damage and mechanical stress.

During stylet probing for a feeding site, herbivores may inadvertently damage cells along the stylet path and release stored plant signals that stimulate expression of PR, *Wfi1,* and *SLW* genes. Alternatively, movement of the stylet between cells, which disrupts essential cell-to-cell contacts, or puncture of the feeding-site cells and consumption of liquids may be perceived as a physical stress generating signals to activate gene expression. Both hydraulic and electrical signals are implicated in the induction of wound-response genes. However, it is unlikely that these mechanical signaling mechanisms can account for the temporal and spatial differences in SLW gene expression induced by the silverleaf and sweetpotato whiteflies.

A component of a herbivore's saliva is likely to provide the general and specific elicitors for *PR, Wfi1,* and *SLW* gene expression, respectively. Liquidimbibing insects, like aphids and whiteflies, secrete two types of saliva along the stylet path and at the feeding site: a rapidly gelling, sheath saliva and a watery, digestive saliva. Salivas have been characterised in a small number of homopteran and hemipteran insects. Sheath salivas are composed primarily of protein, phospholipids, and conjugated carbohydrates. When egested, sheath salivas polymerise around the flexible stylet to form a protective shield, thereby limiting direct contact of the stylet with the plant apoplast. It is not clear whether unpolymerised sheath materials are elicitors in these plantinsect interactions.

The composition of the watery, digestive saliva is more complex and variable containing a wide array of enzymes including pectinases, cellulases, amylases, proteases, lipases, alkaline and acidic phosphatases, and peroxidases. In addition, chitosan is secreted by gall mites at the feeding site. The general and speciesspecific elicitors may correspond to one of the known salivary constituents or may be an uncharacterised component of the saliva. The chitosan (Figure 3), oligogalacturonides created by pectinases (Figure 3), and ROS from peroxidases are known elicitors of wound- and/or defense-signaling pathways (Figures 1, 2).

The general and species-specific elicitors may be directly synthesised by the insect or may be a product of endosymbiotic bacteria. An elicitor could also be generated by concerted biochemical activities of the insect and the plant, similar to the synthesis of volicitin (Figure 3), an inducer of terpenoid volatile production in lepidopteran caterpillars.

PR Proteins, Secondary Metabolites, and Resistance Genes

The role of PR proteins and other herbivore-induced gene products on plant resistance to herbivores that use a piercing/sucking mode of feeding is unknown (Tables 1). The accelerated expression of PR genes in resistant plant-herbivore interactions suggests that one or more components induced by herbivore feeding may function in antixenosis or antibiosis. However, because many PR proteins induced after pathogen infection are active against a subset of pathogens, it is unclear whether any PR proteins influence resistance to piercing/sucking herbivores.

The use of plant mutants that constitutively activate or suppress the JA/ ethylene- and SA-dependent defense pathways will be useful for dissecting the impact of these pathways on resistance to phloem-feeding insects. Examination of transgenic plants that up- or down-regulate specific PR genes may also provide insight into the roles of individual PR proteins in herbivore resistance. Analysis of transgenic plants overexpressing insect and plant chitinases suggests that chitinases have a more limited impact on herbivore interactions.

Nonproteinaceous, secondary metabolites appear to influence phloem-feeding insects more profoundly. For example, volatiles derived from SA (MeSA) and lipids (C_6 volatiles) accumulate in response to aphid feeding and actively deter aphid settling and fecundity, respectively. In addition, several secondary metabolites (that is, acyl-sugars, glucosinolates, and hydroxamic acids) have established roles in resistance to phloem-feeding insects. Genes that control the production of these compounds are being used in breeding programs to enhance insect resistance.

In addition to these quantitative traits, single genes that confer resistance to nonchewing insects have been identified. Some R genes provide a phloem-mediated resistance that deters aphid feeding. Several R genes confer resistance to a single or a small number of aphid biotypes suggesting that "gene-for-gene"-like mechanisms of resistance are also active against phloem-feeding insects. Examples include Nr in lettuce that confers resistance to a single aphid species, Nasonovia ribisnigri, *Sd1* of apple that mediates resistance to two biotypes of the aphid Dysaphis devecta, but not a third biotype, and *Mi1.2* of tomato that confers resistance to the potato aphid (Macrosiphum euphorbiae).

Mi1.2 was the first insect resistance gene cloned and is a member of the leucine zipper, nucleotidebinding, leucine-rich repeat family of R genes that confers resistance to pathogens. The ability of *Mi1.2* to mediate resistance to the potato aphid and the rootknot nematode (Meloidigyne incognita) is intriguing. The facts that resistance to aphids and nematodes develops in seedlings of different ages (I. Kaloshian, personal communication) and that *Mi1.2* causes an HR in response to M. incognita, but not to M. euphorbiae, suggest that *Mi1.2* dual specificity may be complexly regulated.

The parallels to pathogen compatible and incompatible interactions are most compelling in the interactions of the Hessian fly (Mayetiola destructor) and cereals. The Hessian fly does not extensively damage leaves. These larvae appear to secrete substances that stimulate the release of nutrients from cells. Wheat resistance to the Hessian fly is controlled by more than 26 resistance genes, providing resistance to 13 Hessian fly races or biotypes. Similar to pathogen/plant incompatible interactions, Hessian fly resistance is accompanied by an HR surrounding the larvae. Although the temporal and spatial expression of defense-response genes in these interactions are not yet described, there are intensive mapping initiatives to identify and characterise the wheat resistance genes and their cognate avirulence genes from the Hessian fly.

Attacking Insects and Wound Signalling

Phytophagous arthropods that cause extensive tissue damage induce changes in plant gene expression and accumulation of secondary metabolites similar to mechanical wounding. Multiple wound signaling pathways are active in plants (Figure 2). Wound-signaling pathways control the profound changes in plant cell biochemistry that facilitate recovery and healing at the site of injury (Figure 4). In addition, woundresponse proteins may have a direct or indirect role in (1) limiting

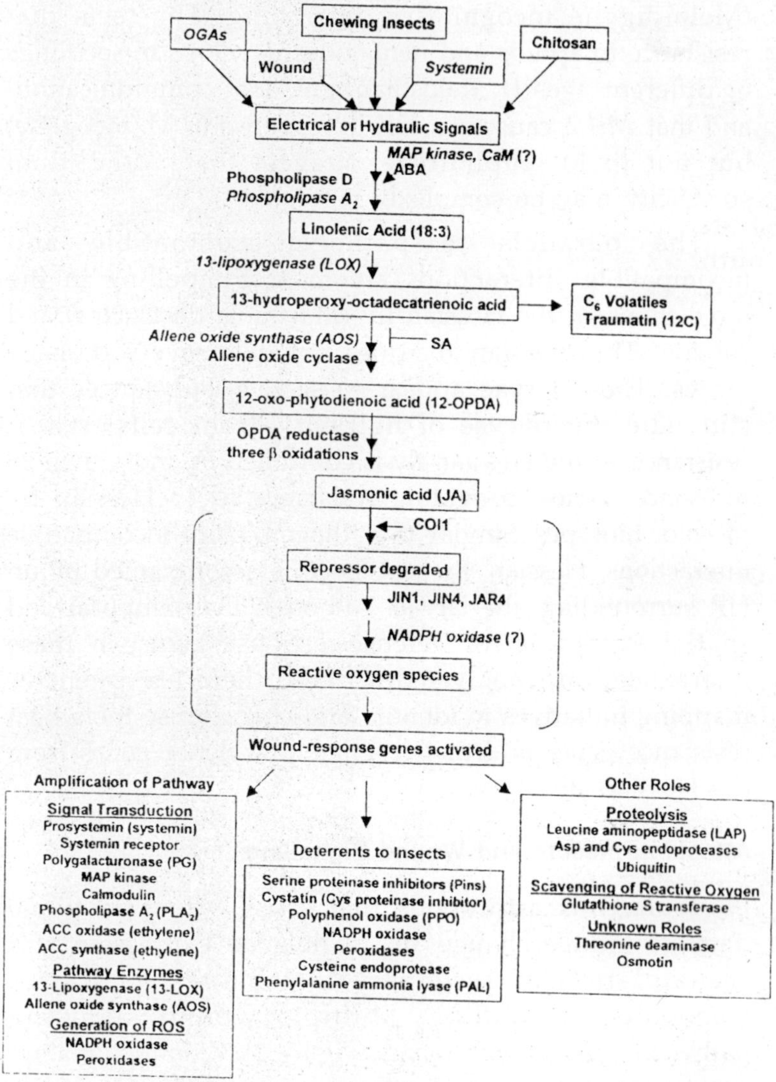

Figure 4. JA-dependent wound-signal transduction pathway of the tomato

damage induced by attacking insects, (2) killing opportunistic pathogens that invade wound sites, and (3) developing an induced resistance (IR) that protects plants from subsequent challenges from pests and pathogens. IR is distinct from the pathogen-induced SAR, but its relationship to the microbe-induced ISR is not understood (Figures 1, 2).

Wound-response genes may be expressed locally or systemically, and these proteins have diverse roles as outlined in Figure 4. Many proteins and secondary metabolites that accumulate after wounding and JA-treatments interfere with insect feeding, oviposition, growth and development, and fecundity or attract herbivore predators. These compounds limit plant injury or restrain insect population expansion (Figure 4). Other wound-response proteins have unidentified roles in the defense and/or tissue recovery.

Octadecanoid Pathway and JA-mediated Wound Responses

Comprehensive reviews detailing the changes in lipid metabolism and mechanisms used to activate JA-mediated wound responses have been published recently. The octadecanoid pathway is induced by herbivores, such as caterpillars and beetles, that chew and tear tissues (Figure 2). Thrips and spider mites, which lacerate cells and imbibe cellular fluids through stylets, also induce JAmediated wound responses. Herbivores not only damage tissues, but their salivary secretions may directly introduce chitosan (Figure 3) and/or polygalacturonase (PG) into the wound site. PG generates oligogalacturonides (Figure 3) from the pectin in the plant cell wall. Both chitosan and OGAs are potent inducers of the Solanaceous octadecanoid pathway (Figure 4). OGAs and chitosan act at the site of release or introduction, respectively; these oligosaccharides are not transported throughout the plant.

Mechanical wounding generates electrical or hydraulic signals that are rapidly propagated from the site of damage (Rhoades and others 1999 and references within). These signals are thought to stimulate the local and systemic release of compounds (OGAs and systemin) that further amplify this signaling cascade throughout the plant (Figures 4, 5A). The increases in plant wound-induced PGs liberate OGAs from pectin. In addition, the bioactive peptide systemin (Figure 3) is produced and transported through the phloem to mediate both local and systemic activation of the octadecanoid pathway. OGA, chitosan, and systemin treatments cause increases in cytosolic calcium, inactivation of H^+-ATPase, membrane depolarisation, K^+ and H^+ fluxes, MAP kinase activity, generation of ROS, and phospholipase A_2 and D activation.

Phospholipases release linolenic acid (18:3) from membranes. LOX converts linolenic acid to a 13-hydroperoxide, which has two possible fates (Figure 4). It is hydrolysed by hydroperoxide lyase (HPL) to generate C6 volatiles and traumatin, a lipid that stimulates wound healing. Alternatively, the 13-hydroperoxide is committed to the octadecanoid pathway by allene oxide synthase (AOS). After six sequential reactions, the bioactive JA is produced (Figure 4). JA and its methyl ester, amino acid, and glucose conjugates are potent signaling molecules. These jasmonates activate wound-response genes by a yet undefined mechanism.

ABA, ethylene, auxin, and SA are additional regulators of the octadecanoid pathway (Figure 3). ABA has an early role in activation of the octadecanoid pathway. The role of ethylene is more complex. Ethylene is essential for JA-mediated wound-response gene expression but antagonises JAinduced nicotine production. Auxin is a negative regulator of wound responses, and the mechanism of auxin action is not understood. SA inhibits both JA synthesis and action (Figure 4).

Lipases also release other fatty acids that are used for the synthesis of additional wound signals: dinoroxo-phytodienoic acid (dn-ODPA) (Figure 3) and dihydroxy(DH)-JA (Figure 3). The octadecanoid pathway enzymes hydrolyze 16:3 lipids to generate dn-ODPA. The production of DH-JA from linoleic acid (18:2) is not completely understood and may vary in different plants. Plants accumulate different levels of JA and its conjugated forms, DH-JA, dn-OPDA, and 12-OPDA (Figure 3). These oxylipins are activators of phytoalexin biosynthesis, alkaloid biosynthesis, or wound-response gene expression in a variety of plants.

At present, it is not known whether the balance of these oxylipin-pathway intermediates and products are different after wounding and insect feeding. Furthermore, it is not known whether insect feeding induces novel fatty acid-derived signals, not present in wounded plants. The recent identification of two, previously undescribed 18-carbon divinyl ether fatty acids, colneleic acid and colnelenic acid, which accumulate in response to Phytophthora infestans infectionindicates that our understanding of the nature of the lipids that accumulate and their roles in defense are still developing.

JA-mediated Wound-response Pathways

Although JA-mediated wound-responses are best characterised, JA-independent wound responses have been implicated in several studies in tomato (Figure 2).

This contrasts to Arabidopsis, where the coordination of the JA-dependent and JA-independent signaling pathways is understood (Figure 5B). At present, wound signaling in Arabidopsis appears to be distinct from tomato. In tomato, a single pathway responds to OGAs, chitosan, and JA (Figures 4, 5A). In Arabidopsis two pathways mediate the spatial and temporal programs of wound-induced gene expression (Figure 5B). A JA-independent signaling pathway is activated by OGAs and chitosan. This oligosaccharide-dependent (OSD) pathway activates wound-response (WR) genes in damaged leaves, consistent with the fact that OGAs are not transported from the site of injury.

The expression of JA-regulated (JR) woundresponse genes is antagonised by OGAs or chitosan in Arabidopsis (Figure 5B). JR RNAs accumulate in apical nonwounded leaves, and at lower levels, in injured leaves. Analysis of mutants that influence ethylene perception or action (*etr1, ein2, and ein3*) show that ethylene antagonises the local expression of JR genes. This is balanced by ethylene increasing levels of allene oxide synthase (AOS), which stimulates the synthesis of JA (Figure 5B).

The OSD- and JA-pathways appear to be reciprocally regulated, because they have opposite responses to kinase and phosphatase inhibitors and internal [Ca^{2+}] fluxes. A comprehensive analysis of genes that respond to wounding and cabbage butterfly larvae (Pieris rapae) shows that these caterpillars induce both JA-dependent and -independent wound-response genes in Arabidopsis. In fact, some Arabidopsis genes are regulated by both pathways. Collectively, data from Arabidopsis indicate that an herbivore will encounter different induced defense molecules in damaged and undamaged leaves. It is unclear whether similar complexity exists in other plants, although it is probable.

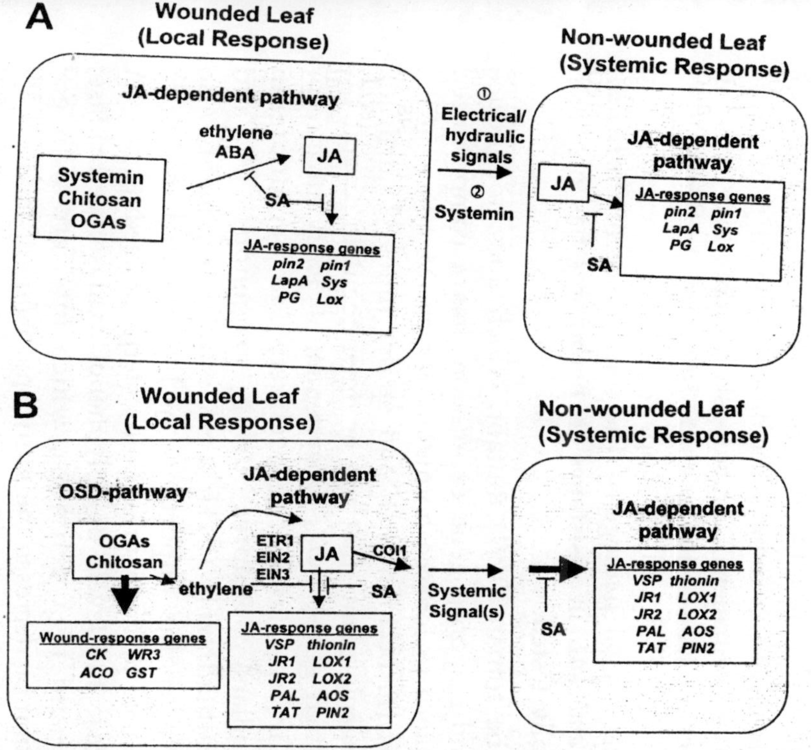

Figure 5. The wound-signaling pathways used in tomato (panel A) *and* Arabidopsis (panel B) *are contrasted and are described in detail in the text*

Octadecanoid Pathway in Herbivore Resistance

Whereas the importance of the JA-independent wound-response in herbivore resistance is not known, activation of the octadecanoid pathway is important for resistance to chewing insects. The Arabidopsis *fad3-2 fad7-2 fad8* mutant abolishes JA and octadecanoid intermediate synthesis and is more susceptible to fungal gnat larvae (Bradysia impatiens). Arabidopsis mutants that have an impact on JA sensitivity (*jin1, jar1, jar4, and coi1*) have been identified and are impaired in their resistance to several fungal pathogens. The importance of *jin1, jin4, jar1, and coi1* in resistance to herbivores has not been formally tested.

The tomato (*def1*) mutant, which does not induce wound-response gene expression, has a compromised resistance to tobacco hornworm larvae (Manduca sexta), thrips (F. occidentalis), and spider mites (T. urticae). Transgenic plants that up- or down-regulate several JA-regulated genes (LOX, prosystemin, or *pin2*) enhance or impair resistance, respectively, to lepidopteran caterpillars. Studies using exogenous JA and benzothiadiazole (BTH; a SA mimic) treatments also emphasise the importance of JA-regulated events in resistance to herbivores. Both caterpillar feeding and JA increase the levels of polyphenol oxidase, peroxidase, *pin2*, and LOX activities, which interfere with gut function or decrease the nutritional value of food.

JA treatments also increase plant tolerance to challenging insects and pathogens and decrease chewing herbivore performance. In a reciprocal fashion, BTH suppresses JA-induced responses, increases susceptibility of plants to beet army worm (Spodoptera exigua), and enhances resistance to pathogens. Analysis of PAL over- and underexpressing lines that stimulate SA-regulated and JAregulated pathways, respectively, supply additional support for the importance of JA-regulated responses in resistance to chewing herbivores.

Herbivore Feeding are not Equivalent

Although many responses to chewing insects overlap with mechanical wounding, these processes are not equivalent. First, herbivore feeding or herbivore regurgitant (contents of the foregut) often cause larger increases in *JA* and wound-response gene RNAs than wounding alone. These data imply that the herbivore oral secretions contain elicitors that stimulate the octadecanoid pathway; however, the nature of these elicitors and their site of action in the octadecanoid pathway are unknown.

Second, the temporal and spatial increases in polyphenol oxidase, peroxidase, pin, and LOX activities after wounding and feeding by different tissue-damaging herbivores (caterpillars, mites, and leaf miners) are distinct. Third, novel genes, which are not induced by wounding alone, are induced by herbivory and tobacco hornworm regurgitant, suggesting novel signaling pathways are induced by chewing insects (Figure 1B).

Fourth, wounding and herbivore feeding provoke the synthesis and release of different sets of volatiles in most plants. Fifth, soybean's woundinduced resistance to herbivores is enhanced on insect feeding or application of regurgitant to plants.

Sixth, insect feeding and regurgitant suppress selected wound-induced responses such as nicotine production. Finally, although wounding of Arabidopsis induces several genes that respond to water-deficit stress, cabbage butterfly larvae do not induce these genes. These observations suggest that caterpillar feeding interferes selectively with wound induction of these genes. Collectively these data indicate that there are substantial differences in plant responses to herbivores and wounding, the magnitude of these differences are not currently appreciated. It is presumed that components of herbivore oral secretions are the elicitors for these herbivore-specific responses.

Volatiles

In response to mechanical or herbivore injury, plants release a complex blend of volatiles providing valuable cues for herbivores and their natural enemies. Volatiles emitted by healthy or infested plants are used by herbivores to discriminate between host and nonhost plants and assess the density of feeding insects on a plant. Volatiles also serve as attractants for herbivore predators and parasites. This indirect defense mechanism provides natural herbivore enemies with reliable, easily detected, and accurate cues to identify plants infested with their host insects. The ability of natural enemies to reduce phytophagous arthropod populations is the basis of biologic control strategies in the field.

Healthy plants release volatiles into the atmosphere, but wounding and herbivore feeding change the volatile blend released by the plant. The constituents of volatile blends are influenced by (1) the herbivore species and its developmental stage; (2) plant species, genotype, and age; and (3) environmental environmental stress. In response to arthropods or mechanical damage, plants release volatiles in two phases. Some volatiles are released immediately (within 1 h) after injury. Labeling studies indicate that other volatiles are newly synthesised (locally and/or systemically) in response to damage and appear 5 to 6 h later.Volatiles have diverse structures and arise from the activities of several biochemical pathways (Figures 6, 7).

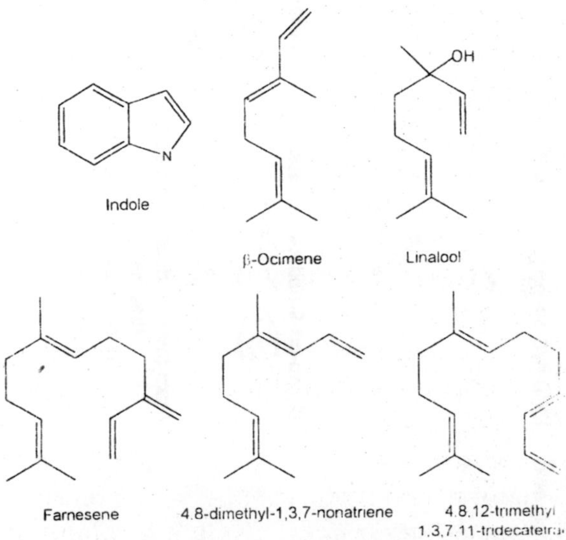

Figure 6. Terpenoid volatiles. The structure for indole and five commonly emitted volatile terpenes are shown.

The most commonly released volatiles include C_6 volatiles (lipoxygenase/ hydroperoxide lyase-dependent pathways), indole and MeSA (the shikimic acid/ tryptophan pathway), cyclic and acyclic terpenoids (isoprenoid pathway), and oximes and nitriles (derived from amino acids). Glucosinolates are partially volatile and are emitted by a limited number of species including the Brassicacae. As with other plant defense responses, wounding and herbivore feeding are not equivalent. In Brassica wounding and herbivore feeding release a similar blend of terpenoid volatiles, but increases in C_6 volatiles and indole glucosinolates are enhanced by insect feeding, relative to wounding.

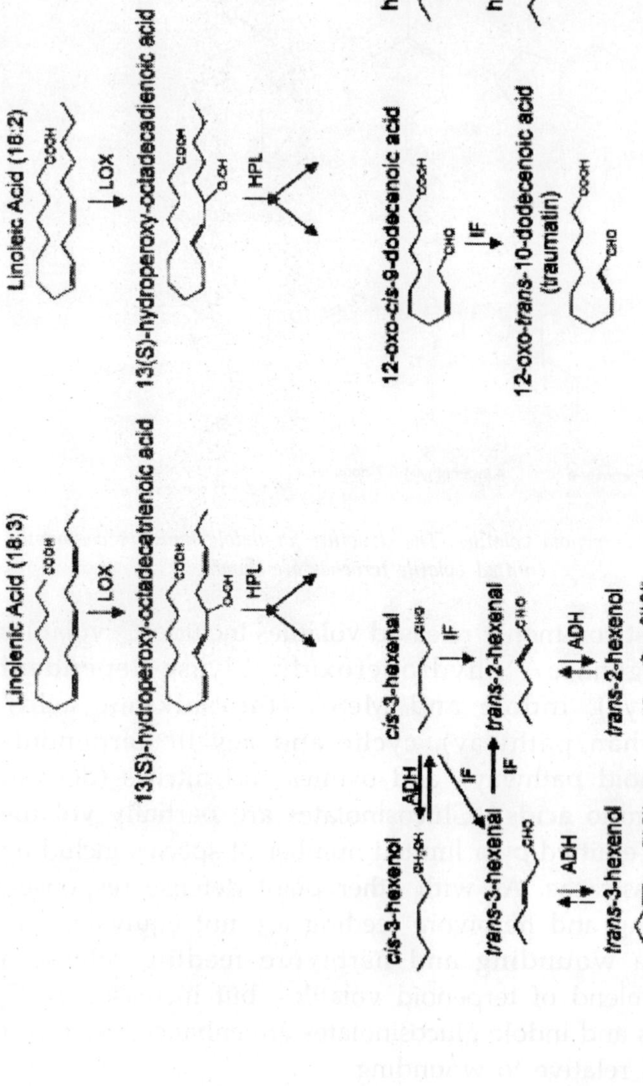

Figure 7. Traumatin and C_6 volatile synthesis

In contrast, maize, soybean, cotton, lima bean, and cucumber emit distinct arrays of terpenoid volatiles after mechanical wounding and herbivore feeding, implicating insect-specific elicitors in the volatile production and release. The volatile blends released by each herbivoreplant interaction are qualitatively and quantitatively distinct. The most abundant terpenoid volatiles are emitted by many different plant-herbivore interactions. It is the molar ratio of the individual volatiles that distinguishes each plant-herbivore interaction.

These quantitative differences are specific cues that either attract or repel natural enemies from herbivore-infested plants. For example, the specialist parasitic wasp Cardiochiles nigriceps is preferentially attracted to plants infested by its host, Heliothis virescens (boll worm), rather to than plants infested with a nonhost insect, Helicoverpa zea (corn ear worm). These observations indicate that the interaction between plants, herbivores, and the herbivore natural enemies is complex and finely tuned.

Volatiles and Phloem-feeding Herbivores

Compared with caterpillars and spider mites, relatively little is known about the volatile blends released by herbivores that do not extensively damage plant tissues. Aphid-infested plants release volatile blends to make them attractive to parasitoid wasps and mediate the density of aphids on plants. The corn leaf aphid (Rhopalosiphum maidis)-maize interaction releases the most common terpenoids emitted after caterpillar and spider mite infestation: b-ocimene, linalool, 4,8-dimethyl-1,3,7-nonatriene, a-farnesene, β-farnesene, 4,8,12-trimethyl-1,3,7,11-tridecatetraene, and acetylated C_6 volatiles (Figures 6, 7).

In addition to terpenoids and C_6 volatiles, MeSA is released after aphid feeding. MeSA may have two roles. MeSA, like SA, activates the SA-dependent defense-signaling pathway and induces PR gene expression and SAR (Figure 1). Although controversial, MeSA may also provide an interplant communication that activates defense responses in neighboring plants. MeSA and the volatile terpenoid b-farnesene are strong aphid repellents (Figures 3, 6). Bernasconi and others have speculated that these compounds allow aphids to avoid settling on plants with high aphid densities. These infested plants have a decreased nutritional quality, attract more parasitic wasps, and have higher induced defenses. All would be conditions detrimental for aphid survival.

JA and Volatile Release

The release of terpenoid and C_6 volatiles are strongly influenced by the release of linolenic and linoleic acids from membranes. These C_{18} fatty acids provide substrates for the synthesis of JA, C_6 volatiles or insectmodified lipid elicitors for volatile production. JA treatments of plants induce release of volatiles not emitted by healthy plants. JA releases a complex blend of terpenoid volatiles from Phaseolus lunatus (lima beans), maize, and Gerbera jamesonii, but the volatiles are not equivalent to those released after herbivore feeding. For example, although the carnivorous mite (Phytoseiulus persimilis) is attracted to volatiles from JA-treated plants relative to healthy plants, they prefer plants infested with their prey T. urticae.

These data indicate that the qualitative and quantitative differences in the volatiles released from mite-infested and JA-treated plants provide important cues to natural spider mite enemies. Finally, in the field, JA-treatment of insect-infested tomato plants enhanced plant-herbivore-herbivore enemy interactions. Tomato plants treated with JA and infested with the beet army worm (Spodoptera exigua) enhanced the attraction and/ or retention of the parasitoid wasp Hyposoter exiguae relative to plants not treated with JA. Nevertheless, a dichotomy exists because wounding does not induce the array of volatiles emitted after JA treatment, but wounding induces JA synthesis.

Wounding may repress the synthesis of the JA- and herbivore-induced volatiles to prevent herbivore enemies from being attracted to noninfested, but mechanically damaged, plants. This suppression could be due to differences in the balance of the oxylipin intermediates and jasmonates that occur after wounding and exogenous JA treatments. Alternatively, repression may be due to mechanisms that are independent on the products of the octadecanoid pathway. Because JA stimulates activities of some enzymes critical for terpenoid volatile biosynthesis, it is possible that the volatile-repression mechanisms may influence expression of volatile biosynthesis genes or regulation of enzymatic activities that control or limit volatile production. The insect-derived elicitors for volatile production may antagonise this suppression mechanism.

C_6 *Volatiles and Herbivore Interactions*

Although large changes in volatile terpenoids occur after herbivore feeding in many plants, changes in C_6 volatiles can also be dramatic and have a strong impact on herbivore-plant interactions. C_6 volatiles are emitted at low levels by healthy plants and are rapidly released in response to insect feeding and mechanical damage. C_6 volatiles, the "green" odors emitted from freshly cut grass, are synthesised from linolenic (18:3) and linoleic (18:2) acids. With the activity of LOX, 13-hydroperoxide lyases (HPLs), alcohol dehydrogenase (ADH), isomerisation factors (IF), and acylases, the C_6 volatiles cis-3-hexenol, trans-2-hexenal, trans-2-hexenol, hexanol, and cis-3-hexenyl acetate are formed (Figure 7).

Table 2. Impact of C_6 Volatiles on Herbivore-plant Interactions

C_6 Volatile[a]	Impact on Herbivores and Plant Gene Expression
trans-2-hexenal	Reduces aphid fecundity
	Reduces spider mite fecundity
	Increases in *AOS, LOX, PAL, VSP, CHS,* and *DFR* RNAs
	Reduces *M. sexta* feeding
cis-3-hexenol	Reduces aphid fecundity
Hexanol	Reduces aphid fecundity
	Increases in *LOX* RNAs
Hexanal	Reduces spider mite fecundity
	Deters *M. sexta* feeding
cis-3-hexenal	Increases in *LOX* RNAs
trans-3-hexenyl acetate	Attractant for the cabbage aphid
C_6 Mixture	Attractant for Colorado potato beetle

Because the levels of 18:3 and 18:2 fatty acids in each organ and plant vary, the balance of the linolenic- and linoleic-derived C_6 volatiles may be distinct and provide important signals in plant-herbivore-herbivore enemy interactions. In some plants, the substrates and enzymes critical for C_6 volatiles are synthesised de novo in response to injury. This is consistent with the fact that wounding and JA enhance phospholipase A_2, HPL, and LOX activities, thereby increasing substrates for C_6 volatile production.

C_6 volatiles influence plant-herbivore and plantpathogen interaction at several levels (Table 2). First, C_6 volatiles stimulate expression of woundresponse genes. It is not presently known whether C_6 volatiles induce a novel set of genes important in herbivore-plant interactions. Second, C_6 volatiles reduce aphid fecundity, spider mite fecundity, and caterpillar feeding . Third, C_6 volatiles are used as attractants for the Colorado potato beetle and specialist aphids. Fourth, C_6 volatiles have antimicrobial and antifungal activity at biologically relevant concentrations. Because defense- and wound-signaling pathways cross talk, deterrence of pathogen growth at the wound site might enhance activity of wound-response pathways critical for wound healing and containment of herbivore damage.

Elicitors of Volatile Production

Two insect elicitors of volatile release have been identified: volicitin and b-glucosidase. Volicitin [N-(17-hydroxylinolenyl)-L-glutamine] (Figure 3) was identified in regurgitant from the beet armyworm and is synthesised by the concerted efforts of the plant and insect. Linolenic acid is supplied by the plant and the glutamine is of insect origin. Application of volicitin to a wound induces the same array of volatiles released by beet armyworm feeding on maize.

β-Glucosidases are the second class of elicitors that mediate volatile release. Treatment of lima beans and B. oleraceae (cabbage) with an almond β-glucosidase releases acyclic terpenes that are typically emitted by the red-spotted spider mite and P. brassicae larvae, respectively. The detection of a β-glucosidase in P. brassicae regurgitant suggests that an insect-derived b-glucosidase may be an elicitor for volatile production. The recent identification of a squash β-glucosidase-like protein gene (SLW3) that is induced by insect feeding suggests that plant-encoded b-glucosidases could potentially play a role in this signaling process.

Regurgitant also contains signaling molecules that enhance expression of wound-response genes, increase JA biosynthesis, and enhance the induced resistance stimulated by insect feeding. It is not presently known whether volicitin, β-glucosidases, or another component of insect regurgitant is responsible for these changes. Relatively little is known about the composition of herbivore regurgitant and less is known about the saliva of chewing insects. The recent identification of a H. zea salivary glucose oxidase suggests that this H_2O_2-generating enzyme could play an important role in ROS production at the site of feeding. ROS are important signals in the plant defense and wound responses.

7

Starch Structure and Digestibility

Commercially, starch is extracted in pure form from a variety of sources. Maize is the predominant source, but wheat, rice, potato, tapioca (maniac) and sago make a significant contribution. Other starch sources include barley, oats, yam, arrowroot and pea. Animal feed contains starch within plant tissues (especially cereal grains) while human foods are often supplemented with starch to provide desirable textural characteristics. The size, shape and granule size distribution of starch reflects the botanical origin as presented in Table 1.

Composition of Starch

Pure starch consists predominantly of a-glucan (ca. 99% of the dry matter) in the form of amylose and amylopectin. Amylose is a roughly linear molecule containing ~99% α-(1-4) and ~1% a-(1-6) bonds with a molecular weight of ~$1x10^5$-$1x10^6$ (Figure 1). Amylopectin (molecular weight ~$1x10^7$-$1x10^9$) is a much larger molecule than amylose and is heavily branched with ~95% a-(1-4) and ~5% a-(1-6) (Figure 1). Each amylose chain is of the order of 1000 glucose units long whereas the unit chains of amylopectin range from ~12 to 120 an hydroglucose units (Figure 1). Starches are defined as waxy when the ratio of amylose to amylopectin is low (~15%), normal when amylose represents ~16-35% and high-amylose (or amylo-) when amylose exceeds ~36%. Waxy mutants are now commonly available for barley, maize and rice and have also been developed recently for wheat and potato. High-amylose mutants, predominantly from maize, are now available commercially.

Table 1. Characteristics of native starch granules from common sources

Starch	Size (μm)	Shape	Distribution
Barley	2-5 (B-granules)	Spherical	Bimodal
	15-25 (A-granules)	Lenticular	
Maize (waxy and normal)	2-30	Spherical/ polyhedral	Unimodal
Amylomaize	2-30	Irregular	Unimodal
Millet	4-12	Polyhedral	Unimodal
Oat	3-10 (single)	Polyhedral	Unimodal
	80 (compound)		
Pea	5-10	Rentiform	Unimodal
Potato	5-100	Lenticular	Unimodal
Rice	3-8 (single)	Polyhedral	Unimodal
	150 (compound)		
Rye	5-10 (B-granules)	Spherical	Bimodal
	10-40 (A-granules)	Lenticular	
Sorghum	5-20	Spherical	Unimodal
Tapioca	5-35	Spherical/ Lenticular	Unimodal
Triticale	1-30	Spherical	Unimodal
Wheat	2-10 (B-granules)	Spherical	Bimodal
	15-35 (A-granules)	Lenticular	

In commercially available purified starches moisture content ranges from 10 to 18% depending on the source, with cereal starches at the lower end and tuber starches at the upper end. Lipid is present in cereal starches only and is positively correlated with amylose content where it may represent ~1.5% of the granule.

Lipid is probably all completed with some of the amylose giving rise to 'lipid free' and 'lipid complexed' amylose. The complication of amylose with lipid affects the structure and functionality of starch granules. Starch lipids (as distinct from surface lipid artefacts) comprise free fatty acids and lysophospholipids (LPL) in different proportions depending on the source of the starch. Purified starches contain low levels of protein (<0.5%) which largely represent the resides of biosynthetic ensymes involved in the synthesis of starch.

Starch Structure

Starch granules are roughly spherical. These granules are synthesised by plants as semi-crystalline matrices, where crystallinity is generated by registration of amylopectin double helices into crystalline lamellae interspersed with amorphous lamellae comprising a-(1-6) branch regions of amylopectin and amylose (Figure 2). In high-amylose starches amylose probably forms double helices and crystalline domains accordingly. In Figure 2 it is apparent that the alternating crystalline-amorphous lamellae provide the basis of semi-crystalline ('dark') growth rings which comprise about 16 crystalline lamellae, which are of approximately the same width as the interspersed amorphous ('light') growth rings (~140nm).

Crystallinity is determined by wide angle X-ray diffraction scattering and is referred to as 'A-type' (found in most cereal starches), as 'B-type' (found in most tuber

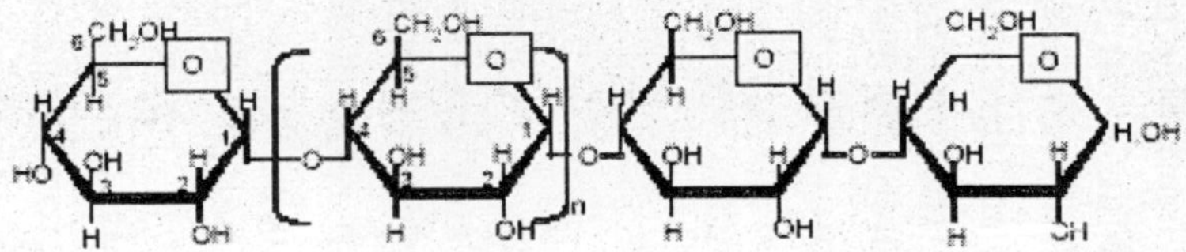

Amylose: α-(1→4)glucan; average n = ca. 1000. The linear molecule may carry a few occasional moderately long chains linked α-(1→6).

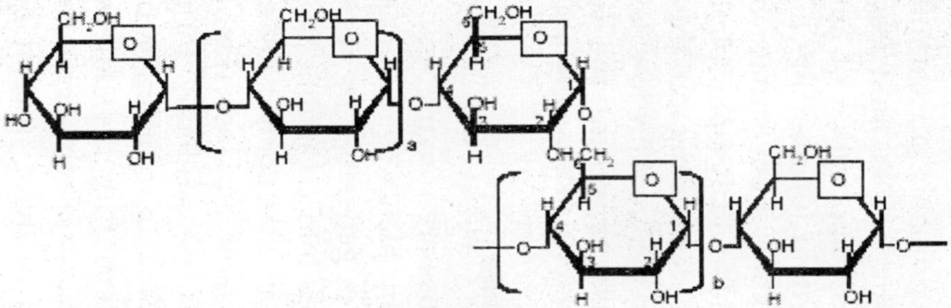

Amylopectin: α-(1→6) branching points. For exterior chains a = ca. 12-23. For interior chains b = ca. 20 -120. Both a and b can vary according to the botanical origin

Figure 1. Basic structure of amylose and amylopectin

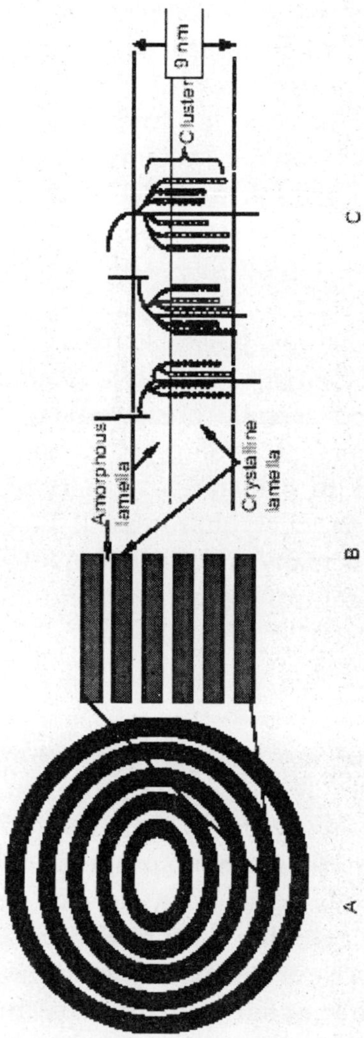

Figure 2. Diagrammatic representation of the lamellar structure of a starch granule. (A) Stacks of microcrystalline lamellae separated by amorphous growth rings. (B) Magnified view of the amorphous and crystalline regions. (C) Double helical structures formed by adjacent chains of amylopectin give rise to crystalline lamellae. Branching points constitute the amorphous regions.

starches), and as C-type' (a mixture of both forms found for example in legume starches). Figure 1 Basic structure of amylose and amylopectin.

Plants synthesise starch in a compact semi-crystalline form to minimise storage space and maximise the concentration of energy stored. The underlying biochemical process is complex and incorporates a number of integrated ensyme systems. Clearly starch provides energy for growth and development when required. Animals consume starch in its native (raw) form. Cereal grains are a good source of energy (referred to as 'starch equivalent') for the majority of animals, including poultry. However, potato starch is not readily digested by monogastric animals. Humans rely exclusively on cooked cereals (bread, pasta, rice) and cooked tubers and roots, as a source of digestible starch. Whilst animals tend to eat starch as part of a plant material (either processed or unprocessed) humans also add starch to food in many modified forms to provide texture (thickener, gelling agent etc.) and inevitably energy (calories).

Gelatinisation

When native starches are heated in excess water the granules undergo a characteristic structural reorganisation which is specific to their botanical origin. They swell and are transformed progressively to an essentially amorphous form with loss of organised structure. Ultimately granule structure is completely lost and a thin paste (<~4%) or gel (>~4%) is formed. At the molecular level the event is initiated by water 'plasticising' amorphous rearrangement and ultimately hydrating double helices as they unravel as a consequence of the elevated temperature.

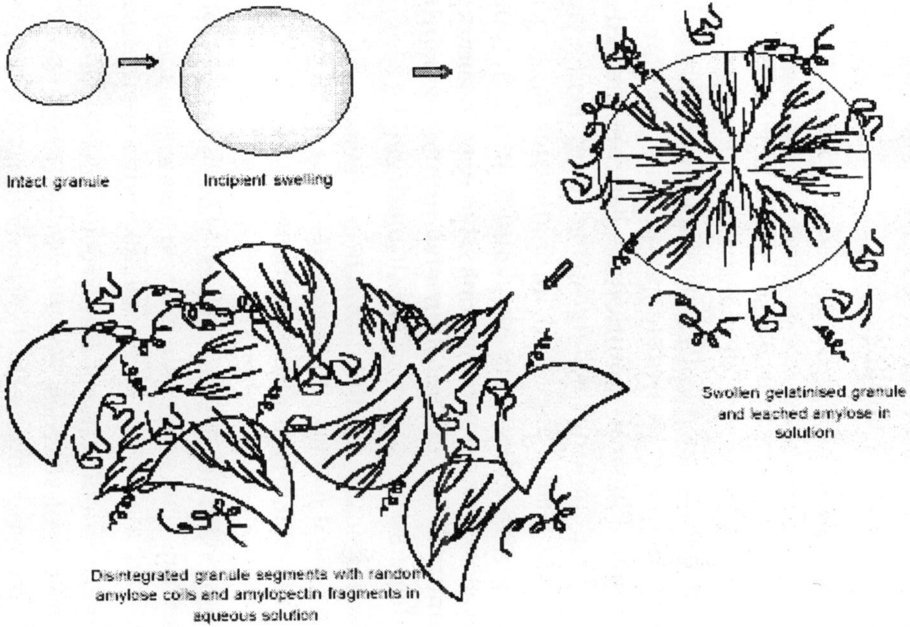

Figure 3. Idealised diagram of the swelling and gelatinisation of a starch granule in the presence of water

Typically this process (in excess water) begins at ~45°C (onset temperature or T_o), peaks at 60°C (T_p) and is complete at 75°C (T_c). This process makes the starch completely digestible by starch hydrolysing ensymes (see below). The gelatinisation process is shown schematically in Figure 3. It should be noted that gelatinisation is restricted if water is limiting.

Modified Starches

Native starches are both indigestible and lack the capacity to contribute to structure in many foods. For this reason they are modified chemically, ensymatically or physically and thus acquire desirable properties for specific applications. Physical modification can be effected by attrition (e.g. milling) whereby the 'damaged' starch granules are easily dispersed in cold water to provide pastes or gels which are digestible. In commercial practice, however, 'pre-gelatinised' starch is obtained by the simultaneous gelatinisation and drying of aqueous starch dispersions (e.g. by roller drum-drying).

This physically modified starch gels instantly on the addition of cold water. Ensymes are used to hydrolyse starches particularly for the production of dextrins and glucose. Although α-amylase and amyloglucosidase are most often used in this respect to make glucose syrup, maltodextrins and crystalline glucose (dextrose), specialist dextrins and maltose may be obtained with α-amylase.

Chemical modifications are used to enhance desirable characteristics of starches. These include 'cross-linking' to provide chemical and thermal resistance and the addition of specific chemical moieties to prevent recrystallisation (retrogradation) of starch gels.

Digestive Enzymes

Animals and humans produce a range of digestive

ensymes which hydrolyse starch. The saliva of both humans and poultry contains α-amylase, which hydrolyses accessible starch, but since this starch remains in the mouth for such a short period, the saliva causes little modification. The pancreas plays a major role in starch digestion where accessible starch in the duodenum receives pancreatic α-amylase into the lumen to hydrolyse α-(1-4) bonds and thus produce glucose sugars, oligosaccharides and dextrins.

Although glucose can be absorbed from the small intestine (especially terminal end of the duodenum and jejunum), maltose and dextrins generated from starch hydrolysis cannot. Absorptive epithelial cells which line the intestinal villi produce a number of other 'brush border' ensymes to allow for digestion and subsequent absorption of such materials (and disaccharides derived from other sources).

These ensymes are not released into the lumen of the small intestine but are bound to the membrane of microvilli. Consequently, some carbohydrate hydrolysis occurs at the surface of epithelial cells in the small intestine rather than the lumen of the gut. Brush border ensymes include sucrose, converting sucrose to glucose and fructose, and lactase converting lactose to glucose and galactose, but these are not relevant to starch metabolism. A trehalase (converts trehalose to two glucose molecules) is also present. However, Maltese (converting maltose generated from α-amylase activity to glucose) and isomaltase (also called a-dextrines) which hydrolyses α-(1-6) bonds of isomaltose and a-dextrins) continue the digestive process and thus convert available starch to glucose. A representation of amylase and debranching ensyme hydrolysis of starch is shown schematically in Figure 4.

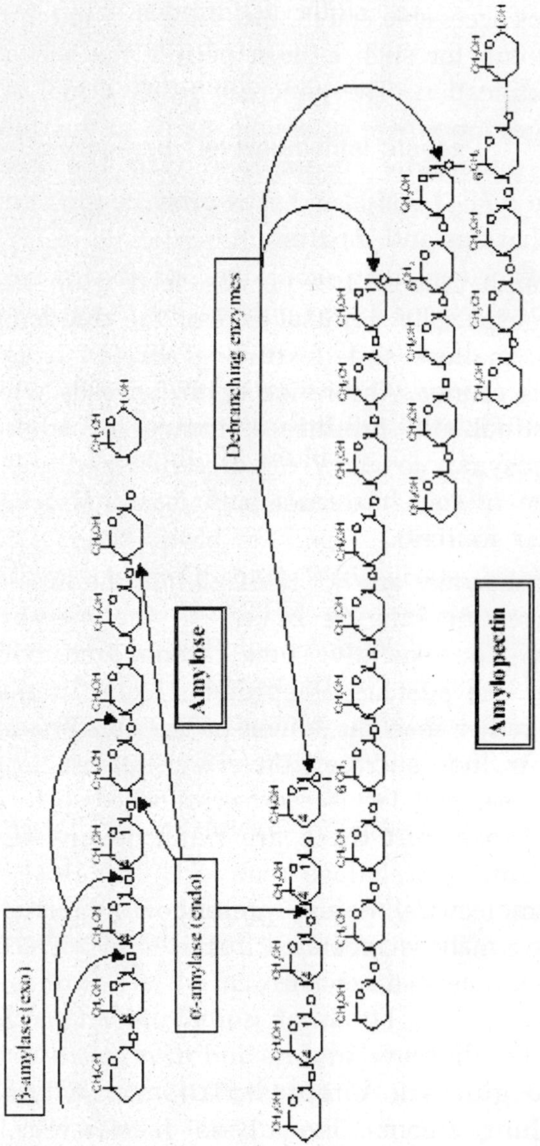

Figure 4. Action pattern of hydrolytic enzymes on amylose and amylopectin

Starch Digestibility

A clear distinction should be made between native granular starch and gelatinised and also mechanically 'damaged starch'. Native granular starch is hydrolysed very slowly by α-amylases and amyloglucosidase. The substrata (starch molecules compacted inside the granules) simply cannot be readily accessed by digestive ensymes and consequently hydrolysis is delayed. In the human digestive system, native starch is one of the four types of 'resistant starch' which may be broadly defined as follows:

— Type I Resistant Starch where starch granules embedded in plant tissue resist digestion,

— Type II Resistant Starch where certain native starch granules like potato and banana are particularly resistant to digestion compared with other starches (e.g. wheat, barley),

— Type III Resistant Starch where starch molecules have recrystallised through hydrogen bonding (as in the stalling of bread),

— Type IV Resistant Starch where novel chemical bonds are formed (other than α-(1-4) or α-(1-6)).

In the human digestive system resistant starches are fermented in the colon and generate short chain fatty acids (SCFA) which can be absorbed and utilised as an energy source. In poultry nutrition, it is similarly recognised that some starches are more resistant to digestion than others. Pea starch (blends) are, for example, more slowly digested than tapioca starch (blends) although researchers in this field often do not define the nature (extent of modification) employed in the feeding trials.

Maize starch granules usually show very strong evidence of α-amylase penetration as pores which are more difficult to detect in other starches. Although α-amylase causes limited hydrolysis of native starch granules, α-amylase is even less effective in hydrolysing starch granules.

It is often reported that the composition or crystalline structure (A-, B- or C-type) of starch granules is the predominant regulator in respect of controlling susceptibility to hydrolytic ensymes. However, it is the view of the present authors that hydrolysis with α-amylase (in model systems) at least is a far more subtle process and controlled by combination of

— the size of the starch granules

— the extent of damage and

— composition and structure

In the digestive tract a further complication is the presence of food/feed structure which limits α-amylase accessibility to starch whether native or modified. Clearly if granules are large they have a smaller surface area to volume ratio and one would predict that amylase binding and potential hydrolysis would be less than for smaller granules as described above. Also, if granules contain more amorphous material (damaged or gelatinised) one would again expect the substrata to be more susceptible to hydrolysis as discussed above.

Regarding granule composition, waxy starches are more crystalline than normal starches and are more readily damaged by milling, making them consequently more easily hydrolysed by α-amylase after the same amount of mechanical damage by milling. However, in native granules there is little difference in susceptibility to ensymes over the short term as the internal structures largely resist enzyme hydrolysis and, since the granules are approximately the same size, there is little capacity for hydrolysis.

For starch to be gelatinised and provide an accessible substrata for α-amylase hydrolysis the gelatinisation process must not be restricted. If water is limiting the amount of gelatinised starch is restricted together with the capacity of α-amylase to hydrolyse the starch. Similarly, the presence of non-starch polysaccharides during gelatinisation restricts the gelatinisation process and subsequently α-amylase hydrolysis. In the digestive tract non-starch polysaccharides as cell wall material prevent starch digestion by preventing the digestive ensymes reaching the starch.

Optimisation of Starch Digestion

For starch to be readily digestible it must be:

— Amorphous (especially physically damaged or gelatinised) not crystalline

— Freely accessible to digestive ensymes (not entrapped in food/feed particles),

— In small particles or preferably solubilised,

— Not associated with other molecules to form associations or complexes (e.g. amyloselipid complexes).

— Not chemically modified in a form that prevents it from acting as a substrata for amylases.

When these criteria are met, the healthy digestive system has a greater opportunity to convert the starch fully to glucose.

8

Regulation of Starch Synthesis in Cassava

Cassava is one of the world's most important, and under-exploited staple food crops and source of income. It is a root crop belonging to section fructicosae of family Euphorbiaceae, Dicotyledonae. It is the third most important source of calories in the tropics consumed by some 600 million people on a daily basis in Africa, Asia, and Latin America. It provides a cheap source of dietary carbohydrate energy (720.1×10^{12} kJ day^{-1}) ranking fourth after rice, sugarcane and maize, and sixth among crops in global production. In Sub-Saharan Africa, it is the leading food security base and a major source of cash incomes for most resource constrained households.

A more significant attribute, however, is that cassava withstands unreliable rains and drought conditions, performs well on soils with low fertility, and has high productivity and low labour demands compared to other major food crops. Moreover, cassava is further more significant in communities with a weakened labour force as a result of the ravages of HIV/AIDs. Consequently, the Food and Agriculture Organisation (FAO) of United Nations (UN) has identified cassava as a crop in East Africa that "will spur rural industrial development and raise incomes for producers, processors and traders, and contribute to the food security of its producing and consuming households".

Recently, the New Partnership for Africa's Development (NEPAD) has identified cassava as 'A Poverty Fighter in Africa' and launched a Pan African Cassava Initiative that seeks to tap the enormous potential of the crop for food security and income generation. In Latin American, the Consortium of Latin America and the Caribbean for cassava research for development (CLAYUCA), a private-public initiative, is promoting cassava as a raw material for animal feed, ethanol and starch industries. In South East Asia, cassava is being exploited for ethanol for automobile fuel. Overall, in a global perspective, there is a commitment to propel the contribution of cassava for human welfare by integrating biodiversity, biotechnology and breeding strategies under the "Global Cassava Improvement Plan (GCPI)".

Together, these initiatives define an evolving and dynamic role of cassava as a catalyst for development. Cassava is a high starch producer with levels between 73.7 and 84.9% of its total storage root dry weight (Fig. 1). This attribute together with the unique properties of its starch creates demand for particular food and non-food applications.

For example, cassava starch readily gelatinises on cooking with water and the solution remains comparatively fluid after cooling. The excellent clarity of its starch is desirable for transparent gels, its bland flavour in pharmaceuticals, and its resistance to shear stress and freezing in the film-forming industries. Besides, cassava starch is used in baby foods, gari, chips, sago, pappads, paints, corrugated boxes, plastics and the tanning of leather. More recently, cassava has found a speciality in the production of synthetic rice.

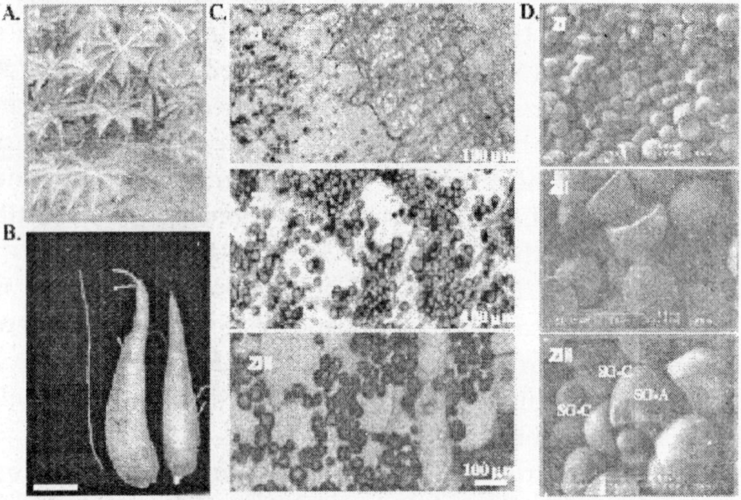

Figure 1. Display of a cassava plant and its starch granule architecture. (a) Cassava plant. (b). Cassava storage roots. (c). Iodine stained map showing cellular distribution of amylose and amylopectin. (d). The scanning electron micrograph of starch granule architecture. Abbreviations: ZI, periderm region; ZII, cortical region; ZIII, parenchyma region; SC-A, starch granule type A; SC-C, starch granule type C; SC-Ca, starch granule type Ca as described (Baguma et

In spite of its potential, cassava starch remains under-exploited mainly because of considerable fluctuations in starch grades and qualities supplied. In response, a project "metabolic engineering of starch biosynthesis in cassava" was developed to contribute to improved starch quantity and quality.

Starch

Starch is the most important form of carbon reserve in photosynthetic eukaryotes or their nonphotosynthetic derivatives (apicomplexa parasites or dinoflagellates) It is a glucose polymer of α-glucans ƒvlinked by α-1,4 bonds and branched at α-1,6 positions. It exists in the leaf chloroplasts as transitory starch (i.e. the primary product of photosynthesis) and in the amyloplasts as storage starch. It is a huge (0.1 > 50 μm in diameter) complex quaternary structure made up of two major glucan polymers: amylopect in (85 - 70%) and amylose (15 - 30%) (Fig. 2). Amylopectin (10^7 - 10^9 Da), by far the major component in leaf starch is composed of intermediate size α-1,4 linked glucans that are clustered together and hooked to longer spacer glucans by α-1,6 linkages, and is responsible for the granular nature of starch.

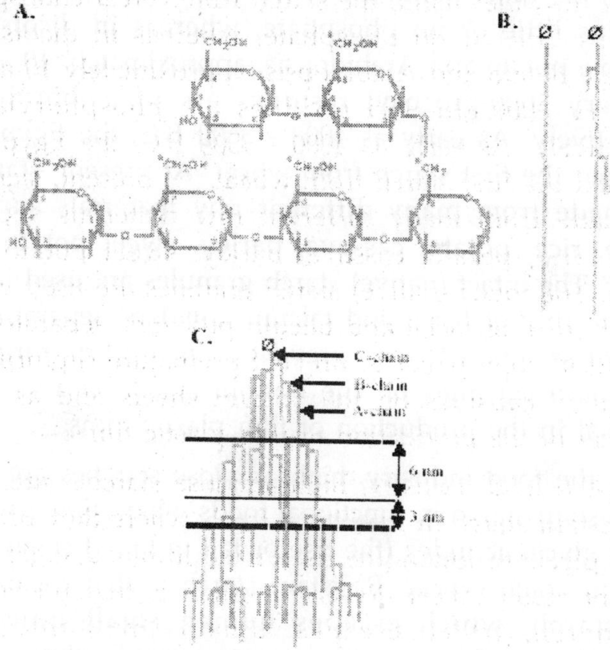

Figure 2. Schematic illustration of starch structure. A. Chains of α-1,4- and α-1,6-linked glucosyl residues. B. Amylose. C. Cluster structure of amylopectin with 3-nm amorphous lamellae (containing the branched regions) and 6-nm semicrystalline lamellae (containing ordered double helices), which alternate with 9-nm periodicity. Φ, reducing end.

Amylose (10^5 - 10^6 Da), which constitutes between 11% and 37% of storage starch, is smaller, essentially linear with less than 1% α-1,6 branches and synthesised within the matrix formed by amylopectin. In general, amylopectin is similar to glycogen (10^7 Da) except for its fewer branch points (ca. 5% of the total linkages), which are discontinuously arranged resulting into clusters of unbranched chains (Fig. 2c). Other quantitatively minor components of starch include proteins, lipids and minerals. Cereal starches have protein levels between 0.25 and 0.5%, while in potato tuber and cassava, the levels are generally <0.1%.

On the other hand, the starch from cereal endosperm contains little or no phosphate, whereas in dicots, for example potato and Arabidopsis, approximately 10 and 1 in every 2000-glucosyl residues are phosphorylated, respectively. As early as 4000 - 3500 B.C, the Egyptians obtained the first starch from wheat. At present, starches are made from many different raw materials such as maize, rice, potato, cassava, barley, sweet potato and wheat. The intact (native) starch granules are used in the manufacture of facial and talcum powders, separators of carbonless copy paper to prevent premature rupturing of ink micro capsules on the bottom sheets and as filler material in the production of thin plastic films.

In the food industry, high amylose starches are used as resistant starch in functional foods where they provide a low glycemic index (the rise or fall in blood sugar) and prevent colon cancer. Resistant starch is that fraction of the starch, which escapes human small intestinal digestion and enters the large bowel where it is fermented by the resident microflora. The resulting short chain fatty acids are metabolised and are beneficial in many health conditions. Paste producing starches are used to thicken products such as soups, sauces, gravies and dairy products.

In addition to the food industry, there are many industrial uses of starch. Starches with high levels of amylopectin are used in paper coatings and adhesives, as viscosity modifying agents in drilling mud during oil exploration, and as joint compounds for finishing gypsum panel walls in the construction industry. In agrochemicals; starches are used as mulches, for pesticide delivery and seed coatings, while in pharmaceuticals, they are used as diluents, binders and in drug delivery.

The synthesis of α-1,4 glucans consists of three critical steps in the chloroplast and amyloplast; the supply of glucose-6-phosphate (Glc-6-P) into the plastid, the synthesis of ADP-glucose (ADPG) from Glc-1-P, and the synthesis of starch from ADPG. Briefly, the first committed and rate-limiting step involves the synthesis of ADPG from Glc-1-P and ATP, catalysed by ADP-glucose pyrophosphorylase (AGPase; EC 2.7.7.23). Once activated, the nucleotide sugar (ADPG) is transferred by the starch synthase (SS; EC 2.4.1.21) to the non-reducing end of an α-1,4 glucan resulting into the generation of linear α-1,4 glucans. Following this, the linear α-1,4 glucans are used as substrates by starch branching enzyme (SBE or Q-enzyme; EC 2.4.1.18) to introduce α-1,6 interlinear chain linkages resulting into amylopectin.

Subsequently, amylopectin is crystallised into starch by the concerted effort of starch debranching enzymes (DBE; EC 2.4.1.41), phosphorylase (P-enzyme; EC: 2.4.1.1) and glucanotransferase.Finally, UDP-glucose: protein glucosyltransferase or amylogenin (38 or 45 kDa, EC 3.6.1 category) has been speculated to be involved in the initial priming process of starch synthesis. In subsequent sections, the role and sub cellular localisation of each of these enzymes is described. Figure 2. Schematic illustration of starch structure. A. Chains of a-1,4- and a-1,6-linked glucosyl residues. B. Amylose. C. Cluster structure of amylopectin with 3-nm amorphous lamellae (containing the branched regions) and 6-nm semicrystalline lamellae (containing ordered double helices), which alternate with 9-nm periodicity., reducing end.

Synthesis of ADPG

ADP-glucose pyrophosphorylase catalyses a rate-limiting reaction in prokaryotic glycogen and plant starch synthesis. In either case, all share similar molecular size, catalytic and allosteric regulatory properties but differ in higher-order protein structure. As for bacterial AGPase (*glgC*), it is a homotetrameric enzyme encoded by a single gene. Conversely, higher plant AGPase is a heterotetrameric enzyme ($\alpha_2\beta_2$) composed of a pair of large subunits and a pair of small subunits, encoded by different genes (D-enzyme, EC 2.4.1.25).

At the regulatory level, small effector molecules whose nature reflects the major carbon assimilatory pathway of the organism modulate the enzyme catalytic activity. Indeed, bacterial AGPase are activated by intermediates of glycolysis [e.g. pyruvate, fructose-6-phosphate (Fru-6-P), fructose-1,6-bisphosphate (Fru-1,6-bisP)] and inhibited by AMP. Likewise, bluegreen algae and higher plants AGPase are activated by 3-phosphoglyceric acid (3-PGA) and inhibited by Pi, key intermediates in CO_2 assimilation by the C_3 pathway. Structure-function analysis of AGPase indicate that the sequences of the small subunits particularly the lysine residue near the C-terminus are highly conserved between species, whereas those of the large subunits are more divergent, suggesting different roles in enzyme function.

Actually, the small sub unit is crucial for catalytic activity while the large sub unit is important for its regulatory properties. In addition, the small sub unit by itself is capable of forming a homotetrameric enzyme exhibiting near-normal catalytic properties but with impended allosteric regulatory properties. In contrast, the large sub unit, is incapable of forming an active enzyme, but enhances the allosteric properties of the small sub unit. Recently, it has been shown through phylogenetic studies that plant AGPase's are distinctly localised and subdivided into leaf, stem, root and endosperm types. At plastid level, AGPase are further classified into two groups based on sub cellular localisation, i.e. plastidial and cytosolic forms.

Extended phylogenetic analysis of the small sub unit proteins suggest that the cytosolic type probably evolved from the leaf localised type. Such lineage raises the question of a modulating factor between the two spatially localised active types: the photosynthetic and sink storage tissue types. In consonance, it has been shown that photosynthetic tissue has a pool of triose phosphates that accumulate in the chloroplast during the light cycle. Likewise, in sink tissue, 3-PGA pool is generated as an intermediate in the respiratory process of glycolysis and may be quite low whereas Pi peaks during starch accumulation.

Taken together, this shows that in sink tissue, Pi inhibition may dominate over 3-PGA activation of AGPase, and phosphate may be the key physiological factor that modulates AGPase activity. This is further corroborated by the fact that maize endosperm AGPase, which is mainly cytosolic, exhibits significant sensitivity to Fru-6-P activation atypical of chloroplast or tuber AGPase. Conversely, tissues without an extra plastidial AGPase, glucose-6-phosphate (Glc-6-P) is transported into the amyloplast, where plastidic phosphoglucomutase converts it to Glc-1-P for ADPG synthesis by plastidic AGPase. Insight into the role of cytosolic and plastidial AGPase in starch synthesis has been gained through studies involving mutants of maize, Shrunken-2 and Brittle-2 and barley, Ris. 16. Ris. 16 lacks cytosolic AGPase activity but has unaffected plastidial activity. Ris. 16 exhibits reduced starch content indicating that a cytosolic AGPase is required to achieve the normal rate of starch synthesis.

On the other hand, both A- and B-type starch granules are present in Ris. 16 showing that the cytosolic form is not necessary for the synthesis of these two granule types. In barley, the plastidial activity by itself is sufficient for normal starch synthesis, albeit reduced rate of accumulation. In contrast, maize plastidial activity alone is not sufficient for normal rate of starch accumulation. Considering that these conflicting lines of evidence are not a consequence of experimental error, it is deduced that knowledge gained from one species may not directly apply to other allied crops.

Synthesis of Amylose

GBSSI belongs to the class of starch synthases. All members appear to share the same basic structure, consisting of a glass domain (substrate-binding site), a typical transit peptide and eight motifs. The class is classified into three distinctly localised fractions in the plastids: those bound exclusively to the granule (granular-bound starch synthases, GBSS); those with exclusive or nearly exclusive activity in the soluble phase (starch synthases, SS); and ones present in both the granule-bound and soluble phase. In addition, the fractions are further subdivided into four subclasses based upon cDNA and amino acid sequences, i.e. GBSS (60 kDa), SSI (57 kDa), SSII (77 kDa) and SSIII (110-140 kDa).

More recently, pea GBSSI has been further subdivided into GBSSIa and GBSSIb isoforms. Likewise, in monocots, there is evidence to suggest that SSII further diverged into two subdivisions, SSIIa and SSIIb. The synthesis of amylose was first attributed to the major granule-bound starch synthase I (GBSSI) many decades ago. Since then, this discovery has been corroborated by several independent studies involving waxy mutants with a defective gbssI gene product. Such mutants have been identified in various species e.g., rice, maize, wheat, barley, amf potato and Iam pea.

To corroborate the waxy phenotype, gbssI in potato was antisensed and analyses of the corresponding amylose content in the transgenic lines showed a marked decrease. Unexpectedly, despite compelling evidence that GBSSI is the sole enzyme in amylose synthesis, graminea waxy mutants were recently shown to accumulate normal starch granules in tissues such as pericarp, leaf, stem and root indicating that another gene(s) controls amylose production. For example, the leaves and stem of waxy rice, the leaves and pericarp of waxy maize, the pericarp of waxy wheat, and the pods, leaves and nodules of Iam pea all contain amylose. This observation raised the possibility that although GBSSI has been considered a sole player in the synthesis of amylose, other players are involved. Recently, this was strengthened by the isolation of a second isoform of GBSSI, designated GBSSII from waxy wheat and pea leaves, designated GBSSIb. Indeed, both isoforms were demonstrated to be involved in the synthesis of amylose.

Synthesis of Amylopectin

Various studies have demonstrated that SSI, SSII and SSIII are involved inamylopectin synthesis, although the role of SSI and SSIIb remains unclear. In pea, studies of rug5, which is closely allied with a defect in ssIIgene showed altered amylopectin branching pattern with decreased intermediate--sized glucans (dp 15 - 25) and increased short-chain glucans (dp < 10). In potato tubers, antisense inhibition of SSII and SSIII singly or in combination resulted into a significant shift from longer to shorter chains. In wheat, elimination of the ssIIa gene product (SGP-1) manifested a phenotype with reduced starch content and altered starch structure.

Likewise, the fairly recently discovered sugary-2-like phenotype in maize mutants defective in SSIIa, demonstrate that SSIIa is involved in starch synthesis. Overall, present understanding indicate that loss of SSII (dicots) or SSIIa (monocots) results in reduced starch content, reduced amylopectin chain length distribution, deformation of the starch granules, altered physicochemical properties of starch and perturbed crystallisation. It has also been shown through studies of the maize mutant, dull 1, Chlamydomonas reinhardtii mutant, STA3, and transgenic potato carrying an antisense SSIII construct that SSIII contributes to amylopectin branch-length distribution.

Starch Branching Enzymes

Starch branching enzymes (SBE, 40 - 152 kDa) are involved in amylopectin synthesis. They catalyse the hydrolysis of α-1,4 linkage and subsequent formation of α-1,6 glucosidic bond between the cleaved chain and a hydroxyl group on C^6 of a glucosyl moiety of an α-1,4 glucan template. They belong to the α-amylase family characterised by a catalytic (β/α)8-barrel domain. The domain encompasses specific active sites that arise from the inter-connecting β-loops providing for substrate binding and catalytic activity. To date, two classes have been identified based on amino acid sequences and in vitro catalytic properties of purified enzymes.

In monocots, SBEII have further been subdivided into SBEIIa and SBEIIb filiations depending on specific catalytic properties, length of amino acid residues in the N-terminal domain and C-terminal polyglutamic acid repeats. Type I SBE, has been identified in maize, rice, pea, cassava, and wheat. More recently, a novel SBEI cDNA, sbeIc, was isolated from developing endosperm of wheat, uncovering the wide diversity of type I SBEs. Type II SBE, has been identified in pea, maize rice, barley, wheat, cassava and sorghum. Notably, SBEI and SBEII isoforms exhibit different substrate preferences.

SBEII isoforms have lower affinity for amylose than SBEI, implying that SBEI isoforms uses longer glucan chains than SBEII isoforms. Also, it has been found that SBEI isoforms have greater branching activity and preferentially use amylose as a substrate. Moreover, their protein structure exhibits further contrasting architectural features, i.e. SBEII possess an extended serine rich N-terminal domain while SBEI have a lengthy C-terminus of ca. 100 amino acid residues. The N- and C-terminal domains have been associated with protein anchoring and regulatory functions, respectively.

It has been shown from several studies that SBE isoforms are differentially and independently expressed during organ/tissue development and within the amyloplast. Genes encoding SBEI are commonly and constitutively expressed in photosynthetic and vegetative tissues while SBEII are preferentially expressed in starch storage compartments. The reverse is rare but common. For example, potato SBEII is predominantly expressed in leaves with very low but detectable levels in the tuber, whereas SBEI is the major isoform in the tuber. This disparity is widened by the expression pattern of sbeII in monocots like in wheat, barley, maize and rice where maximal expression of sbeII is attained early in kernel maturation.

By contrast, SBEI is strongly expressed in the second half of embryo development in maize, wheat and barley, while maize SBEIIa is more highly expressed in leaves than in endosperm. In pea embryo, both forms are present at comparable levels in the soluble fraction, whereas in potato, SBEI is the most abundant soluble fraction than SBEII. Most significantly, whereas both SBEIIa and SBEIIb share similar patterns of action, similar expression profiles, and are both distributed between the granule and stroma, only mutations in SBEIIb cause a high amylose phenotype in cereal grains. This raises the question of function for the different isoforms. In elucidating the function of the different SBE isoforms, insight has been gained from analysis of starches derived from contrasting sbe mutants.

In both monocots and dicots, mutational and gene suppression of SBEI cause minimal effects on general starch synthesis and composition in tubers, leaves and endosperm. However, it has been shown that loss of BEI protein in rice results into significant changes in the fine structure of amylopectin and physicochemical properties of the resulting starch in the rice endosperm. As for SBEII, exclusive elimination of SBEII in potato led to increased levels of amylose, although, combined suppression of both SBEII and SBEI markedly increased the amylose content of the resulting starch phenotype.

This showed that the seemingly obscure function of SBEI in starch synthesis might be reflective of overlapping specificities and complementation between SBEI and SBEII (SBEIIa and SBEIIb).

It has also been proposed that the enzyme might not interact with the substrate until SBEII (SBEIIa and SBEIIb) have acted. In spite of these efforts, the question of function for the different isoforms remains not clearly resolved. Notably, isoform substrate specificities (cleavage and branching junctions), configuration of the substrate molecules (helical structure or interchain) and enzyme complexes during starch synthesis require to be examined.

Debranching Enzymes (DBE)

Starch debranching enzymes belong to the α-amylase super family. They hydrolyse the α-1,6 glucan branches of amylopectin. The group is subdivided into two classes. The direct DBE, involved in the hydrolysis of α-1,6-linkages of α-polyglucans and the indirect DBE, engaged in hydrolysis of α-1,6-branches by 4-a -glucanotransferase and amylo-1,6-glucosidase. Direct DBE are further subdivided into pullulanase-type or R-enzyme (EC: 3.2.1.41) and isoamylase (EC: 3.2.1.68). The defining difference is their substrate specificity in which pullulanases debranchpullulan and amylopectin but not glycogen, whereas isoamylase debranch bothglycogen and amylopectin.

Moreover, pullulanases generates maltosyl groups, while isoamylase releases maltotriosyls and largeoligosaccharides. In cereals, isoamylase is a larger (400 kDa) multimeric enzyme composed of one type of isoamylase subunit. Conversely, in potato, two distinct subunits define an equally large heteromultimeric enzyme. There is good accumulated evidence that DBEs plays a crucial role in starchbiosynthesis. This has been accumulated through the analysis of sugary-1 mutants of maize and rice endosperm, sta7 mutant of Chlamydomonas, dbe1 mutant in Arabidopsis, which accumulate a water-soluble polysaccharide (WSP), designated phytoglycogen.

This analysis led to the trimming model, which asserts that the synthesis of amylopectin, its organisation and incorporation into a starch granule is the result of 'trimming' by DBE of the highly branched glucans synthesised by SS and SBE. This model hypothesises discontinuous synthesis by way of preamylopect in, a theoretical intermediate of amylopectin, through the activity of DBEs that results in altered efficiency of the enzyme complex. In spite of this account, the trimming model has some limitations as observed in Arabidopsis dbe1 mutants, which accumulate starch and phytoglycogen suggesting that the accumulation of phytoglycogen in wild-type is presumably forestalled by the action of a specific isoamylase enzyme, perhaps in concert with other glucan-degrading enzymes.

To date, analysis of the Arabidopsis genome has revealed that there are three genes encoding isoamylase like proteins (ISA1, ISA2, and ISA3). These are conserved in divergent plants and evidence from potato and Arabidopsis indicate that the protein encoded by ISA1 and ISA2 are subunits of one heteromultimeric isoamylase protein in vivo. Loss of ISA1 and ISA2 in Arabidopsis manifest a phytoglycogen accumulating phenotype suggesting that both are essential in starch synthesis. This leaves ISA3 with undefined role but presumably might be important in starch degradation.

Taking this knowledge into consideration and from the point of view that during mobilisation of transitory starch, both starch synthesis and degradation occurs concurrently, it seems plausible that ISA3 is required to debranch glucan structures that arise during starch degradation, while ISA1 and ISA2 are essential for amylopectin synthesis. Taken together, the latter prompted the development of a new model "simultaneous processing", where specific isoamylase, in this case ISA1 and ISA2, clears the stroma of WSPs generated by ISA3 and other starch degrading enzymes, leading to simultaneous stalled phytoglycogen accumulation but increased amylopectin synthesis. In spite of these exciting models, it remains unresolved whether glucan trimming, WSP clearing or some other mechanism explains the mode of action of DBE.

Disproportionating Enzyme (D-enzyme)

D-enzyme was first reported in potato tuber. Later, it was detected in beans, carrot, peas, spinach, tomato and Arabidopsis. The enzyme disproportionate soluble oligosaccharides of at least three glucose residues or amylopectin into maltooligosaccharides. In vitro analysis of potato D-enzyme showed that the protein is capable of transferring branched glucans and or producing cyclic glucans. An attempt to understand its function in starch biosynthesis through mutation studies in Chlamydomonas and Arabidopsis showed conflicting roles.

In Chlamydomonas, the mutants showed significant reductions in total glucopolysaccharide, abnormal starch granule, altered component proportions and atypical amylopectin chain-length distribution relative to the wild type. Conversely, Arabidopsis mutants overproduced starch, while transgenic potato tubers transformed with antisense D-enzyme constructs showed no effect on starch synthesis and its fine structure. These inconclusive findings suggest additional research to clearly establish the function of D-enzymes in starch synthesis.

RI protein has been reported to be involved in phosphorylation of tuber starch. Recently, ADP-glucose pyrophosphatase, an enzyme likely with dual role in controlling levels of ADPG linked to starch synthesis and other metabolic pathways has been added to the list of enzymes involved in starch biosynthesis. Also, starch phosphorylase (P-enzyme) has been shown to be involved in the catalysis of reversible phosphorylytic cleavage of starch and it seems very likely that P-enzyme is a part of the complex degrading enzymes involved in starch breakdown. It has also been shown that several enzymes involved in starch degradation e.g. α-amylase, β-amylase, D-enzymes, α-glucosidase (maltase), glucan water dikinase (GWD) and α-glucan phosphorylase are equally important players in starch biosynthesis.

Most significantly, is the fact that glucan water dikinase (GWD) appears to control the overall rate of starch breakdown with a central rate limiting role in starch breakdown machinery and downstream starch synthesis. A detailed account on the roles, sub cellular localisation and regulation of these enzymes has been recently reviewed. In the same review, the authors present a model for starch breakdown and its connection to starch synthesis. Lastly, sucrose invertase (SI), sucrose synthase (SuSy) and sucrose phosphate synthase (SPS) are crucial in sucrose metabolism and contribute to sink strength in divergent crop species.

Amylogenin

Despite the intense research in plant carbohydrate

metabolism, several intriguing questions still remain. One relates to the primer for starch synthesis, if indeed such a thing exists! The rift in our understanding is further aggravated by the fact that the mechanism underlying the priming of glycogen synthesis in E. coli is equally unresolved. Nonetheless, glycogenin has been identified as the priming protein for the initiation of glycogen synthesis in yeast and mammalian cells, but the situation in plants remains a mystery. Along with this evidence, it is expected that protein similar to glycogenin may be involved in the initiation of starch synthesis in plants.

A selfglycosylating protein named UDP-glucose: protein glucosyltransferase or amylogenin (38 kDa) first reported in potato and later in maize has been thought to be involved in starch synthesis. One of the main reasons for assuming that amylogenin functions in starch synthesis was its occurrence in starch synthesising endosperm tissue of maize. However, recent results indicate that this may not be true. The evidence include the covalent nature of amylogenin-glucan glycosidic linkage (β-glucosylarginine instead of α-glucosyltyrosine), its substrate specificity (UDPG and not ADPG) and its sub cellular localisation (Golgi and not plastid) bringing into question its function in starch priming, but not necessarily against its function in polysaccharide synthesis.

So, what is amylogenin? It has been revealed through polypeptide functional studies and tryptic peptide sequence matches that amylogenin is closely related to Pisum sativum reversibly glycosylated polypeptide (PsRGP1), which is said to be involved in the synthesis of xyloglucan. Notably, RGP1 does not interact with ADPG, the principle substrate for starch synthesis in plants. Moreover, RGP1 is localised in the trans-cisternae of

Golgi dictyosomes and not plastids, the sites for xyloglucan and starch synthesis, respectively.

Furthermore, the fact that amylogenin associates with starch synthesising maize endosperm tissue, does not exclude the possibility that as this tissue cellularises from the liquid endosperm, it also synthesises cell wall polysaccharides. Thus, neither the substrate specificity nor the localisation of amylogenin seems to agree with a role in starch priming. The question of what primes starch synthesis still remains raising the concern of what could be the other possible primer alternatives? In response, two possibilities have been suggested. First, short maltooligosaccharides (MOS) may prime amylose synthesis. Second, amylopectin-primed amylose synthesis has equally been identified as a plausible model.

The evidence for these models has been accumulated from MOS accumulating mutants and pulse-chase studies involving either a pulse of ADP[^{14}C]Glc to isolated starch granules or $^{14}CO_2$ to intact plants, followed by a chase period in unlabelled substrate, for the former and latter, respectively. Furthering our understanding of the mechanism(s) that govern the initiation of starch synthesis is likely to have profound impact on the starch industry.

Starch Modification

Prospects for manipulating starch composition using modern molecular tools have given an impetus to research on starch metabolism. Transgenic plants and starch mutants such as *ae, amf, sug, wx, shr1, shr2* and *flo* harbouring altered starch content and composition have been identified and or produced in several plants such as barley, maize, pea, rice, oats, potato and wheat, opening such a possibility in cassava and kindling the need to understand the key regulatory steps controlling starch biosynthesis or mobilisation in its storage roots.

In dicots, examples of genetically engineered crops include potato with a 30% increase in starch yield derived from altered AGPase activity. In addition, an attempt has been made to increase the yield of starch through heterologous expression of E. coli PPi in transgenic potato. Interestingly, the extent of modification depended upon the potato variety used. In the Russell Burbank variety, the modification produced a 30% increase in starch content, but in the Prairie, such changes were not recorded. This disparity suggested that there might be a limit to the amount of starch, that potato can accumulate, which the Prairie variety had already reached, or the result was reflective of pleiotropic effects of the modified AGPase on other enzymes in potato.

Aside modifying AGPase by itself to enhance the rate of starch synthesis, new elegant strategies based on enzyme-substrate kinetics have been designed with spectacular outcome. For instance, increasing the levels of a 'plastidic ATP- ADP transporter', a protein involved in supplying ATP for the reaction, increased the yield of starch in potato. Other notable modifications include antisense SBEI in transgenic potato and simultaneous alterations of SS activity also in potato. The latter affected starch content, glucan chain lengths, granule morphology and properties of the starch.

More recently, simultaneous antisense inhibition of the genes for three starch synthases in transgenic potato yielded starch with short-chain amylopectin and very little amylose. In monocots, *Himalaya 292*-a novel barley cultivar—derived from a single nucleotide change in ssIIa, has higher amylose (70%) and NSP than the parent strain. As for oats, 17 transgenic lines with reduced amylose content have been produced by antisense inhibition of gbssI. In wheat, altered activity of sbe led to higher proportion of A-type granules in transgenic lines. Also, in wheat loss of one or two of its gbssI gene, remarkably reduce amylose content.

More recently, *notch2* isoamylase deficient mutants of barley associated with an apparent increase in phytoglycogen. For maize, altered AGPase activity led to 18% increase in starch yield. Also, su1 maize isoamylase mutants accumulate phytoglycogen instead of amylopectin. In rice, EM557 mutant lacking the BEI protein altered the fine structure of amylopectin. Likewise, reduction in ISO1 protein activity in rice endosperm markedly altered amylopectin into a water-insoluble modified amylopectin and a water-soluble polyglucan. Beyond the test tube and other laboratory conditions, it is recognised that gene action is influenced by other factors including diurnal functions.

For instance, in rice it has been shown that temperature at which a plant develops influence the proportions of amylose and amylopectin produced by affecting the activity of different SSs. Furthermore, there is evidence that altering the rate at which starch is produced affects the fraction composition. At lower rates of production a greater proportion of the starch is converted into amylopectin, because of higher enzyme units per substrate molecule. This raises the concern that attempts to improve starch yield might result in starch with increased amylose content, which would be undesirable for some applications.

While there are significant species-dependent differences among starches, starch in its native form has limited applications. As a consequence, it is often necessary to chemically or physically modify the starches to tailor for various uses. However, if the plant that produced the starch could carry out such modification in planta that would be of great advantage. Thus, the quest for in planta modifications of starch synthesis has gained momentum. Examples of such in planta modifications include starch with increased degree of phosphorylation, high or all-amylose starch, all-amylopectin starch, and starch with defined granule sizes. Despite the progress made in analysing and manipulating the mechanism of starch synthesis in plants, the interactive roles of the key enzymes involved and their metabolic regulation are not clearly understood.

For example, rug pea mutant originally intended only to affect starch production, also resulted in wrinkled seeds containing abnormal quantities of lipid and storage proteins. Such unintended occurrences alludes to one cardinal fact that prior to any attempt aimed at generating novel starch, it would be important to make sure that the intended modification has no or negligible effects on other untargeted nutritional qualities. Finally, one recurring theme while dissecting starch metabolic pathways is that the enzymes involved come in multiple forms, which differ in their physical and chemical properties and in the type of starch that they produce. While this increases the complexity of the process, it nonetheless increases the scope for metabolic engineering.

Taken together, the aforementioned aspects in allied crops convincingly demonstrate that modification of starch biosynthesis in cassava is equally possible. Moreover, many of the genes involved in its starch biosynthesis have been cloned and insight into their regulatory properties has been obtained. Thus, prospects for carefully planned modifications are equally possible in cassava.

Regulation of Starch Biosynthesis

Regulation of starch synthesis involves a flexible but complex network of highly interactive catabolic and anabolic reactions in the cytosol and plastids. Most interestingly, the plant cells must know when to use sucrose for starch synthesis and or for other metabolic processes. Also, it is well established that different genes and their isoforms diversely located in the plant cells are involved in starch metabolism. Thus, it is important that their activities are well coordinated consistent with their roles.

Overall, available evidence suggest that the entire regulatory network for starch synthesis relies on genetic (including developmental cues), cellular metabolic status and environmental factors for its function. Major regulatory mechanisms of starch synthesis are described below.

Developmental Regulation

The role of developmental regulation of starch synthesis genes has been established through spatial and temporal analysis studies in a number of crop species. It is governed by the interplay of genetic control, mitotic activity, histo differentiation and cellular metabolic status. In barley, faba bean, wheat and maize; it has been shown that the pattern of storage starch accumulation correlates well with cell expansion but is spatially distinct from the pattern of mitotic activity. Detail analysis has identified that metabolic status maintains a specific stage of differentiation and or directs the next developmental programme including starch biosynthesis.

Most remarkably, it has been shown that the expression of storage-associated genes is dependent on changes in sucrose levels in a his to differentiation-dependent manner with specific-isoform expression profiles. Consistent with this, sbeII and sbeI genes in pea are expressed early and late in embryo development, respectively. Similarly, in maize endosperm, it has been shown that sbeI is strongly expressed between 10 and 28 dap whereas, sbeIIb is expressed throughout endosperm development. In contrast, rice sbeI is immediately detectable post-pollination, while sbeIIb in rice is detectable at 5 dap with maximal expression between 5 and 15 dap.

In sorghum, the onset of sbeIIb and sbeI expression starts from 10 dap, attains maximum expression between 16 and 22 dap, and declines thereafter. A similar pattern for sbeI in wheat has been reported. In barley, sbeIIa and sbeIIb predominate during early stages of grain development, while sbeI is maximally expressed during late development. Taken together, these findings suggest that the expression profile and expression levels of different sbe isoforms is a well-coordinated process that involves the interactive effects of cellular mitotic changes, metabolic signals and genetic control, although the whole mechanism remains to be clearly understood.

Diurnal Regulation

Plant metabolic networks involve different gene expression-mediated pathways. These expressions manifest as steady-state levels or as day/night recurring cycles called diurnal rhythms. Diurnal rhythms are entrained by light and revolve on an internal circadian clock. Light mediates circadian function through signalling pathway(s) that involve the phytochrome systems, which upon light reception are transported to the nucleus. In the nucleus, the phytochrome interacts with *PIF3 (PHYTOCHROME INTERACTING FACTOR 3)* a basic helix-loop-helix transcription factor and binds to the G-box element. The effects of light results in recurring responses with some exhibiting typical overt circadian rhythms.

The basic model of a circadian system is composed of three control components: input, oscillator and output. The input, which is triggered by light and temperature, transmits a signal to the oscillator for resetting. In turn, the cycling of the input pathway trains the clock to a specific schedule sustained by the oscillator. In a coordinated sequence, the oscillator controls the output to overt circadian rhythms. To date, there are indications that starch metabolism and some of the genes involved are entrained to the circadian rhythm of the plant.

Light-responsive elements; G, GT1, and GATA have been identified in promoters of light regulated starch synthesis genes and convincingly shown to be essential for light responsive transcription. Cis-acting elements that bind specific transcription factors have also been identified in the gbssI promoter of Snapdragon and sbe2-1 in Arabidopsis. Besides, diurnal oscillation of AGPase in Chlamydomonas and gbssI in Snapdragon suggest involvement of circadian mode of regulation.

At the level of starch, it has been shown that plants transferred from a typical diurnal regimen to continuous light perturb starch synthesis during subjective night, and in one case, starch was even degraded. More recently, work involving micro-array analysis has revealed that several Arabidopsis genes encoding starch synthesis/degrading enzymes [e.g. chloroplastic isoforms of α-amylase, β-amylase, glucan water dikinase (GWD), sbe2-1] are under the influence of circadian clock. However, there exist conflicting evidence with regard to the relationship between gene expressions, protein level and enzymatic activities of such diurnally regulated genes. For instance, it is not known in the case of α- and β-amylases whether the amounts of protein or enzymatic activities match with gene expression profiles and levels.

In all cases, however, entrainment to the circadian mode of regulation may serve to prime the events in starch metabolism along with other key factors such as day length, temperature and nutrient availability. Whatever the significance, little is known about the underlying molecular mechanisms.

Sugar Signalling

Sugar signalling is a well-established phenomenon in a number of plants. Sugar signalling involves three critical steps i.e. sugar/signal sensing, signal transduction, and action at target genes. Sugar sensing might arise from changes in apoplast sugar concentrations, sugar flux through the plasma membrane, fluctuations in intracellular sugar levels, and, from glycolytic, plastidial and or vacuolar sugarspecies inter conversions. Whatever the mode of sensing, it is established that sugar signals mediate many metabolic processes in plants including germination, seedling growth, leaf and root development, starch synthesis and senescence, in concert with expression of the relevant genes. It has been shown that, cab and rbcs (encoding chlorophyll a/b-binding protein and the small sub unit of rubisco, respectively) are repressed by sucrose and glucose, while starch synthesisgenes and the patatin promoter are activated by sucrose.

Additionally, the expression of α-amylase is repressed by sucrose. In spite of this evidence, it remains unclear how the sugar effects are sensed and or transduced to the target gene(s), although plant HXK's have been identified and implicated in the process. It has been shown that plant HXK's associates with chloroplast, amyloplast, mitochondrial membranes, or with the golgi apparatus providing further evidence for a role in starch synthesis. Additional evidence indicates a role of fructokinases and the existence of a glycolysis-dependent pathway. Recently, the existence of a sucrose sensor has been reported. It is speculated that the sensor functions either as an active sucrose carrier or by a metabolic-derived signal.

In the first case, the transporting carrier activates the next protein in the signalling pathway. In the second case, metabolites downstream, act as metabolic messengers. In congruence, transporting and none transporting sucrose sensors have been identified. That said, the mode of sucrose transport and the nature of sucrose transporters remain highly controversial. Nonetheless, changes in extracellular pH and membrane potential are implicated. Here, it involves activation at the plasma membrane H^+-ATPase, which in turn generates a negative membrane potential and acidifies the extracellular environment relative to the cytoplasm.

Most interestingly, sucrose transporters display interactive specificity with the glucosyl hydroxls 3, 4 and 6 of sucrose unravelling the significance of side group orientation or linkage position, and its binding with sucrose is dependent on the presence of a fructosyl moiety for a hydrophobic surface. Further downstream, sugar signals and or sensors, initiate complex signalling networks to target genes interlinked by plant Ser/Thr protein related kinases, SnRKs. Specific cis-elements e.g. sugar responsive elements.

Hormonal Signalling

The interactive effects of sugar and hormonal signalling

have long been known to modulate plant metabolic processes, including but not limited to seed maturation, dormancy and changes in gene expression. It has been shown that abscisic acid (ABA) enhances sugar induction of Apl3 and sbe2.2 gene expression in Arabidopsis. In addition, it has been shown that some Arabidopsis ABA-insensitive mutants (*sis*5, sugar insensitive; *sun*6, sucrose uncoupled; gin6, glucose insensitive = *abi*4) and ABA synthesis mutants (*sis*4, *gin*1 = *aba*2) are equally defective in sugar sensing, and vice verse.

Additionally, ABA-insensitive mutants (*abi*1, *abi*2 and *abi*3), unaltered in sugar sensitivity exist, indicating that a specific ABA-related transduction pathway mediates sugar sensing. Nonetheless, it is established that many of the sugar-sensing mutants are altered in their ability to synthesis or transduce ABA signals. ABA signalling involves putative ABA receptors (extracellular or intracellular), cell-surface membrane proteins including ion channels, glycoproteins and membrane trafficking proteins, secondary messengers such as phosphatidic acid, Ca^{2} and protein phosphorylation/de-phosphorylation cyclic cascades. The opposition of kinase-catalysed phosphorylation and phosphatase hydrolytic de-phosphorylation maintains the steady state phosphorylation levels of most enzymes in cyclic cascades.

PP2 subgroup is further subdivided into PP2A, 2B, and 2C based on whether or not its activity is dependent on divalent cations. PP2B and PP2C activity requires Ca^{2+} and Mg^{2+}, respectively, but PP2A, like PP1 does not. Sequence and structural analyses indicate that PP1, PP2A, and PP2B are tightly related and defined as PPP, whereas PP2C, pyruvate de-hydrogenase phosphatase and several Mg^{2+}-dependent Ser/Thr phosphatases form a distinct family, named PPM. Most notably, it has been revealed from Arabidopsis sequence database and other genome-wide surveys that several genes encode for proteins related to Ser/Thr phosphatases.

At least, 7 genes encode PP1 catalytic sub unit, 6 for PP2A catalytic sub unit, 70 for PP2C and a handful for other PPP types. Taken together, their effects depend on the manner by which they are regulated. To date, it is well established that multiple regulatory mechanisms are involved

(i) at the level of gene expression,

(ii) at the level of protein localisation,

(iii) at the level of substrate specificity, and

(iv) through regulation by the activity of other enzymes.

For instance, genes for PP2A catalytic subunits are constitutively expressed but their expression levels are developmentally regulated. In addition, the noncatalytic domains of PP functions to localise catalytic activity in the cell while their interactive effects with other regulatory proteins modulate their activity. In this case, it has been clearly established that a bicyclic cascade involving phosphorylase kinase and phosphoprotein phosphatase-1, enzymes that are involved in the glycogen phosphorylase bicyclic cascade, modulate the activity of glycogen synthase. Lastly, it has been shown that PP2A and its regulatory sub unit are crucial for plant developmental processes and hormonal signal transduction. For instance, the rcn1 gene, which encodes a 65 kDa A-sub unit of PP2A, is essential for normal root development in Arabidopsis.

Expression Profiles of Starch Synthesis Genes

Previously, cassava cDNA's encoding AGPase (both small

and large subunits), GBSSI, GBSSII and SBEI were cloned Munyikwa et al.,. Here, a partial cDNA encoding starch branching enzyme II (SBEII) was cloned, apparently derived from a single-copy sbeII gene. Database comparisons of its nucleotide sequences with genebank sbe sequences showed a high degree of similarity (<79%) to other plant sbeII. It is weakly and constitutively expressed in all tissues during early development, but markedly expressed in the storage root at maturity. This finding indicated that sbeII is expressed in dynamic patterns in developing cassava storage roots. These results differed from its botanical relative potato where sbeII is more strongly expressed in the leaf than in the tuber. Also, it was found that sbeII transcription activity increased starting 90 days after planting (dap) and attained maximum abundance at 360 dap.

A similar shift in spatial and temporal expression profile was observed for sbeI in this study. The concerted expression profiles of sbe uncovered the existence of a well-coordinated developmental programme that ensures a high transcriptional activity of starch biosynthesis genes in the mature storage root. In addition, the finding that sbe expression and SBE activity prepatterned storage root development together with cassava's phasic development corroborated this finding. Another key finding in this investigation was that the 360 dap old storage roots had gbssII and gbssI transcripts, as well as a longer gbssI transcript, designated gbssI'. The gbssI' transcript was identified as a variant transcript of gbssI characterized by the retention of the first three introns.

Developmental and Diurnal Regulation in SBE Expression

The transcription and enzymatic activity of sbe prepatterned the storage root development states, but the spatial and temporal developmental starch accumulation patterns remained unclear. Following this, a study was designed to examine the relation between the storage root histo differentiation stages, and sbe transcription and endogenous enzymatic branching activity. At 180 dap, sbe transcription activity increased from the periphery to the hilum of the storage root. Interestingly, at 360 dap, greater sbe transcription activity relocated away from the hilum to the periphery.

Likewise, the same pattern was seen for endogenous enzymatic branching activity, revealing the physiological relevance of the relocation. Developmentally, these findings agreed with the storage root compartmentalisation growth model and provided additional evidence that the pattern of sbe transcription activity and its enzymatic activity prepatterns starch accumulation. Furthermore sbe transcription activity remained spatially distinct from the pattern of mitotic activity, since no coincidence was observed between maximum sbe expression and cellular mitotic activity as deduced from cellular density and size.

To further corroborate this evidence, histological studies were performed to map the in situ starch granule organisation in the storage root, and establish a relationship between starch granule organisation and mitotic gradient. The map showed a gradual increase in both cellular and starch granule size (5 - 40 μm) away from the periphery of the periderm to the hilum inter phased by characteristic secondary xylem starch-depleted buffer zones. Similar results were obtained from granule scanning electron microscopy of granule architecture.

Most notably, the iodine stained monographs showed an inward increment in the relative amylopectin content revealing that the storage root of cassava is not a homogenous tissue. In practice, these results uncovered the existence of distinct starch domains in the storage root of cassava and identified the need to unravel the underlying developmental regulatory mechanism. Developmental regulation is a multi factor regulatory mechanism that in part depends on the prevailing diurnal functions.

Consequently, the relationship between sbe transcription activity and diurnal functions was examined. Maximal expression levels of sbe occurred within 6 h inside the light zone irrespective of the light-dark interface schedules. Although, it remained unclear whether the observed oscillation was typical of an endogenous oscillator or otherwise. To dissect this question, changes in sbe transcript abundance in storage root cells were monitored under prolonged light and or dark conditions. This analysis showed that sbe mRNA oscillated in periods of 12 h within a circadian day.

In turn, this prompted an investigation to establish whether the observed expression pattern was reflective of the endogenous SBE enzymatic activity. The results matched, indicating a well-coordinated biological phenomenon. This phenomenon was taken to mean that the overt oscillations in the transcription activity of sbe are under the regulation of a bona fide semi-diurnal regulatory mechanism, driven by an endogenous semidian oscillator, which is premised in the storage root cells of cassava, although its nature and function remains unknown. At the same time, these results suggested that the semidian oscillator might have evolved to synchronize the photosynthetic apparatus with starch biosynthesis machinery, analogous to the circadian rhythm of cab genes, which prepares the predawn photosynthetic cell for the approaching light exposure.

Interplay between Sugar and Hormonal Signalling Networks

The interplay between sugar and hormonal signalling networks is known to regulate several metabolic processes during plant growth and development. Thus, by considering previously established developmental dependent sbe expression profiles and expression levels, it was equally assumed that sugar signalling plays a role in the regulatory mechanism of sbe expression in the storage roots of cassava. Subsequently, the effects of exogenously supplied sugars on the expression of sbe were studied in metabolically active starch synthesising 360 d-old plants as described.

The transcription activity of sbe was strongly induced by sucrose to levels observed in 360-d-old plants. Glucose and fructose had minimal effects. Surprisingly but significantly, turanose a transportable but non-metabolisable sucrose analog substantially induced sbe expression, suggesting that sucrose stimulation might be independent of its metabolism into constituent hexoses. On the other hand, poorly metabolisable and or non-metabolisable sugars (i.e. mannose, 3-O-methyl glucose, mannitol and palatinose) did not induce sbe expression, suggesting that sbe transcription is independent of osmotic effects.

Most notably, the effects of the two sucrose (Glc[1>2]Fru) analogs {i.e. turanose (Glc[1>3]Fru) and palatinose (Glc[1>6] Fru)} indicated that the stimulatory effect of turanose might have a relation with its transportation into the cell and/or the steric position of hydrogen at positions 2 and 3 in the Fru moiety is important for sbe stimulation. Also, this finding suggested that the Fru moiety must be part of a disaccharide to trigger expression since; Fru by itself was not sufficient to induce readily detectable sbe expression.

Consistent with this hypothesis, it was shown previously that induction by turanose might be associated with its transport across the plasma membrane and the disaccharide-linked fructosyl unit has been identified as essential for a hydrophobic interaction between sucrose and the sucrose transporter, SUT.

Further analysis of the effects of palatinose and turanose were not sufficient to convincingly neither show nor preclude the likelihood of extracellular and or intracellular sucrose sensing mechanism. To dissect this possibility, an attempt was made to unravel the effects of sucrose and turanose by simultaneously examining their stimulatory capacity in the presence or absence of glucosamine, a potent hexokinase (HXK) inhibitor. In this analysis, the finding that sucrose and turanose-induced sbe expression was impaired (sucrose) or abolished (turanose) in the presence of glucosamine suggested that HXK and downstream metabolism of sucrose might be involved.

Broadly, these results were considered to suggest that turanose and or sucrose elicited signal actuates endogenous HXK and in turn, potentiated HXK modulates semidian sbe transcription oscillation. Taken together, these findings implied the existence of an intracellular disaccharide sensor and concomitant recruitment of downstream hexose metabolites in the regulatory mechanism of sbe. Moreover, the existence of multiple sugar signalling mechanisms involving sucrose as a specific signal, signalling at the SUT level and involvement of glycolysis-dependent sensing pathway have been reported previously in a number of crop species suggesting a likelihood of a similar phenomenon in this case. In practice, these results were interpreted to suggest that sbe preferentially recognises a photosynthetic product, sucrose.

In proof of concept, sucrose therapeutic studies were conducted in decapitated plants and we found that sbe transcript levels were strongly enhanced. This result convincingly revealed the requisite for sucrose flux from the photosynthetic source to the sink storage root cells to trigger expression. Furthermore, these data uncovered the existence of both apoplastic (storage root fed) and symplastic (therapeutic fed) modes of sucrose phloem uploading in the growing cassava storage roots.

In potato tuber, phloem unloading is predominantly apoplastic during stol on elongation, but primarily symplastic during the initial phases of tuberisation. In contrast, because cassava storage root concurrently initiates new deposition sites and fill existing ones in a compartmentalised manner, it is possible that apoplastic and symplastic modes of phloem uploading associate with initiation and filling events, respectively. Further analysis showed that sbe semidian transcript and enzymatic activity oscillation in the isolated discs persisted without any damping through day 4, although in decapitated plants, the rhythmicity was abolished. This comparison between the effects arising from intact and uncoupled storage roots suggested that factors in addition to diurnal function and sugar signalling influence the transcription activity of sbe genes.

To investigate this clue, it was speculated that hormonal-effect is involved in downstream signalling cascades for sbe expression. ABA induced sbe expression, strongly enhanced the sucrose-effect, and uncoupled the function of the semidian oscillator. Also, it was shown that ABA effect very likely is independent of sucrose stimulation since by itself and or in perturbed sucrose signalling background, its ability to induce sbe was retained. As a consequence of these results, the role of other regulatory components on nuclear expression of sbe was studied to gain additional insight into the functioning of the entire regulatory network.

Role of Plastid De Novo Protein Synthesis and Protein De-phosphorylation

Data presented previously provided evidence for the role of plastid signalling in eliciting gene expression in the nucleus. As a preliminary approach, a study was performed to investigate whether a similar signalling pathway plays a role in starch synthesis in the amyloplast of cassava. Specifically, the role of plastid de novo protein synthesis and protein De-phosphorylation on sbe transcription activity was investigated. Clearly, the expression of sbe in cycloheximide-treated storage root discs was not affected, indicating that cytosolic de novo protein synthesis is not essential for its expression. In contrast, sbe expression was completely abolished in chloramphenicol-treated discs suggesting that plastid de novo protein synthesis plays a significant role in its regulatory network.

Collectively, these results provided the first evidence that nucleoplastidic signalling might be involved in the regulation of sbe expression. Indeed, nucleoplastidic signalling, commonly referred to as retrograde signalling has been demonstrated previously in a number of plant signalling systems. The mechanism involves the participation of a number of cytoplasmic regulatory factors that link nuclear sensing to plastid signals in the signalling transduction highway. To explore this possibility, the link between plastidial signalling and cytoplasmic protein De-phosphorylation was examined.

Here, okadaic acid, which preferentially inhibits type 1 and 2A protein phosphatases, PP1 and PP2A respectively, by mimicking the role of cytosolic phosphoprotein phosphatase inhibitor 1 and 2, abolished the transcriptional activity of sbe, in the presence or absence of sucrose. Conversely, cantharidin, a potent inhibitor of PP2A, did not, indicating that PP1 is essential for sbe expression. In addition, inhibition of PPI by okadaic acid was considered to suggest that it had delinked nucleocytoplasmic traffic and prevented upstream plastidic signal transduction resulting into abolished sbe expression.

Hence, on the basis of these results, a nucleo-plastid signalling working model for the regulation of sbe was proposed. In this model, sucrose signalling via the plastid (amyloplast) was identified as a major route in the storage root of cassava. It was also found that cytosolic and nuclear signalling pathways alone were not sufficient to elicit sbe expression without the plastid factor. Likewise, the plastid factor by itself was equally not sufficient to elicit sbe expression without the presence of the cytosolic component. Moreover, these findings further revealed that the plastid factor appeared to stimulate nuclear transcription of sbe, possibly after it had actuated relevant cytoplasmic regulatory components.

In addition, the fact that sucrosechloramphenicol-treated discs resulted in abolished sbe expression but that expression was rescued when the discs were treated simultaneously with ABA showed that the sucrose-stimulated plastid signal functions to activate downstream cytoplasmic signalling components. In spite of these intriguing results, the nature of the signal generated by the amyloplast, the signal transduction mechanism across the amyloplast membrane, and the mechanism by which the transduced signal is sensed by cytoplasmic regulatory components remain a mystery.

However, within the wisdom of these data it was speculated that redox signalling might be involved.

Analysis of the regulatory mechanism of the catalytic sub unit of AGPase in potato tubers also indicates that this is the case. Tiessen and co-workers showed that when sucrose was supplied to potato tubers, starch synthesis increased, albeit with low levels of phosphorylated intermediates, including 3-PGA. Similarly, these data demonstrate that sucrose and or its plastid-derived signal is crucial for eliciting nuclear expression of sbe.

9

Sterol Metabolism

Sterols are important for the physiology of eukaryotic organisms. They form part of the cellular membrane where they modulate their fluidity and function and participate as secondary messengers in developmental signalling. Different organisms utilize different sterols. The most important ones are cholesterol, phytosterols, and some steroid hormones in animals, and campesterol, sitosterol and stigmasterol in plants.

Whereas cholesterol is the unique sterol of vertebrates and ergosterol the major sterol of most fungi and some unicellular algae, most higher plants contain a complex mixture in which cholesterol is a minor component and 24-ethyl sterols (sitosterol and stigmasterol) account for more than 60% and 24-methyl sterols account for less than 40%. Configuration at C-24 is 100% R in sitosterol. In contrast 24-methyl sterols are a mixture of 24-epimers where campesterol (24R-methyl cholesterol) and dihydrobrassicasterol (24S-methyl cholesterol) are in variable proportions.

Important differences may occur in some plant families, for instance many plants belonging to the order of caryophillales contain large amounts of Δ7-sterols, spinach (*Spinacea oleracea*) and *Chenopodium rubrum* contain almost only Δ^7-sterols such as spinasterol or 5a–stigmast-7-ene-3β–ol. Likewise, changes in the side chain structure may occur in some plant families. For instance sterols with a double bond at position C-25 occur often in cucurbitaceae and verbenaceae. The sterol composition of brassicaceae to which Arabidopsis belongs, presents a small difference to most plant species for an additional sterol, brassicasterol (ergosta-5,22E-diene-3β–ol), is found in the 24-methyl sterol fraction. It is noteworthy that insects, which constitute more than 80% of animal species, do not synthesize sterols. They use sterols from their diet.

For instance, phytophageous insects are able to demethylate sitosterol into cholesterol, which is finally used to make ecdysteroids involved during embryonic and larval development. Sterols have multiple functions. First of all they are membrane constituents. They accumulate in the plasma membrane where they are present in concentrations similar to those of phospholipids. Several data obtained with model systems show that sterols in condensing the membrane bilayer regulate membrane fluidity and permeability.

In addition to this role played by the bulk of sterols, a regulatory function has been assigned to sterols in various systems :

i) covalent binding of cholesterol to HEDGEHOG is involved in embryonic development of vertebrates;

ii) interaction of cholesterol with caveolin induces the formation of membrane microdomains (caveolae), which may be signaling centers for multiple pathways;

iii) transport of plant sterols by elicitins from *Phytophtora* spp leads to a hypersensitive like response in tobacco. These processes trigger signaling pathways involved in cell division, development or resistance to pathogens. Finally sterols are precursors of compounds with a high physiological activity such as brassinosteroids, an important class of hormones involved in higher plant growth and development, ecdysteroids in invertebrates, antheridiol and oogoniols, sexual pheromones of aquatic fungi belonging to the saprolegniale class. Therefore, sterols, in addition to their own biological activity, are important interfaces of plant insect interactions.

In spite of this recent knowledge, several important questions remain open especially in plants:

i) higher plants synthesize phospholipids containing polyunsaturated fatty acids and bulky 24-ethyl sterols, which are both incorporated into membranes in which they should interact. Is the biosynthesis of both class of compounds co-regulated and, if so, how does such a coregulation work?

ii) Considering the regulatory processes enumerated above, what is the role of sterols in signal reception and transduction ?

iii) It is well established that brassinosteroid biosynthesis constitutes one branch deriving from phytosterol biosynthesis. Whereas sterols are present usually in plants at concentrations in the range of several mg per g of dry weight, a few ng per g of dry weight of brassinosteroids are generally found. Therefore a very small fraction of the phytosterols synthesized by the plant will be used for brassinosteroid biosynthesis. This leads to the existence of a tight control at the junction point between both pathways.

In order to answer these questions, a strategy of gain or loss of function has been applied recently to the sterol biosynthetic pathway. The use in the past of potent and specific inhibitors gave unvaluable informations which in some case were not reliable because of side effects of inhibitors. Recently the impressive development of molecular genetics, the availability of DNA tagged mutants, multiple possibilities offered by stable transformation in sense and antisense orientation and gene silencing, assorted with the sequencing of the Arabidopsis genome led to an important breakthrough in the field of sterol metabolism and functions.

Before entering into this field, a short recall of the state of the art of the sterol biosynthesis is needed. Sterol biosynthesis is complex and involves at least 25 steps from isopentenyl diphosphate, the commited precursor of all isoprenoids, to end pathway sterols. The biosynthetic scheme of Figure 1 is common to most higher plants and also applies to Arabidopsis.

From isopentenyl diphosphate to 2,3-oxidosqualene (OS) the biosynthetic pathway is the same in eukaryotes, however profound differences exist downstream of OS (Figure 2). Whereas OS is cyclized to lanosterol in nonphotosynthetic eukaryotes, it is cyclized to cycloartenol, an isomer of lanosterol possessing a cyclopropane ring in place of the Δ^8 double bond, in photosynthetic eukaryotes.

As end pathway plant sterols do not contain a cyclopropane ring, an enzyme capable of opening it exist in photosynthetic eukaryotes. A third difference consists in the presence in higher plant sterols (sitosterol, stigmasterol...) of an extra 24-ethyl group, resulting from two methylation steps situated between cycloartenol and end pathway sterols. Sterols from vertebrates are not alkylated at C-24, whereas sterols from most fungi

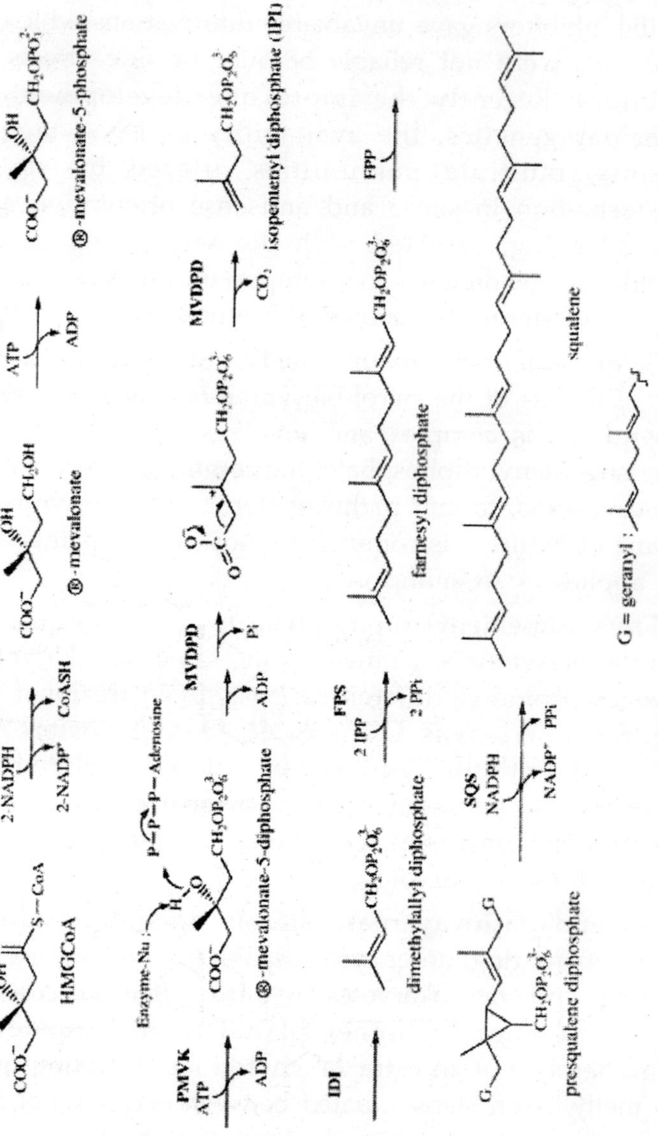

Figure 1. Biosynthetic pathway leading to plant sterols

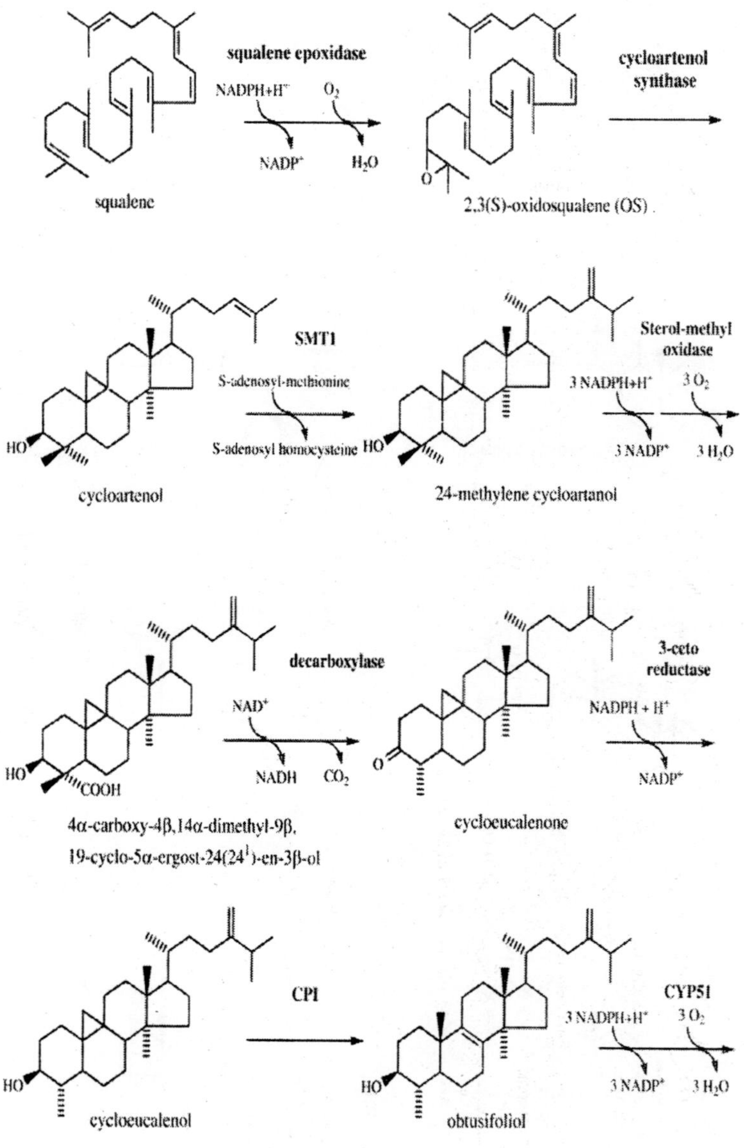

Figure 2. Biosynthetic pathway leading to plant sterols

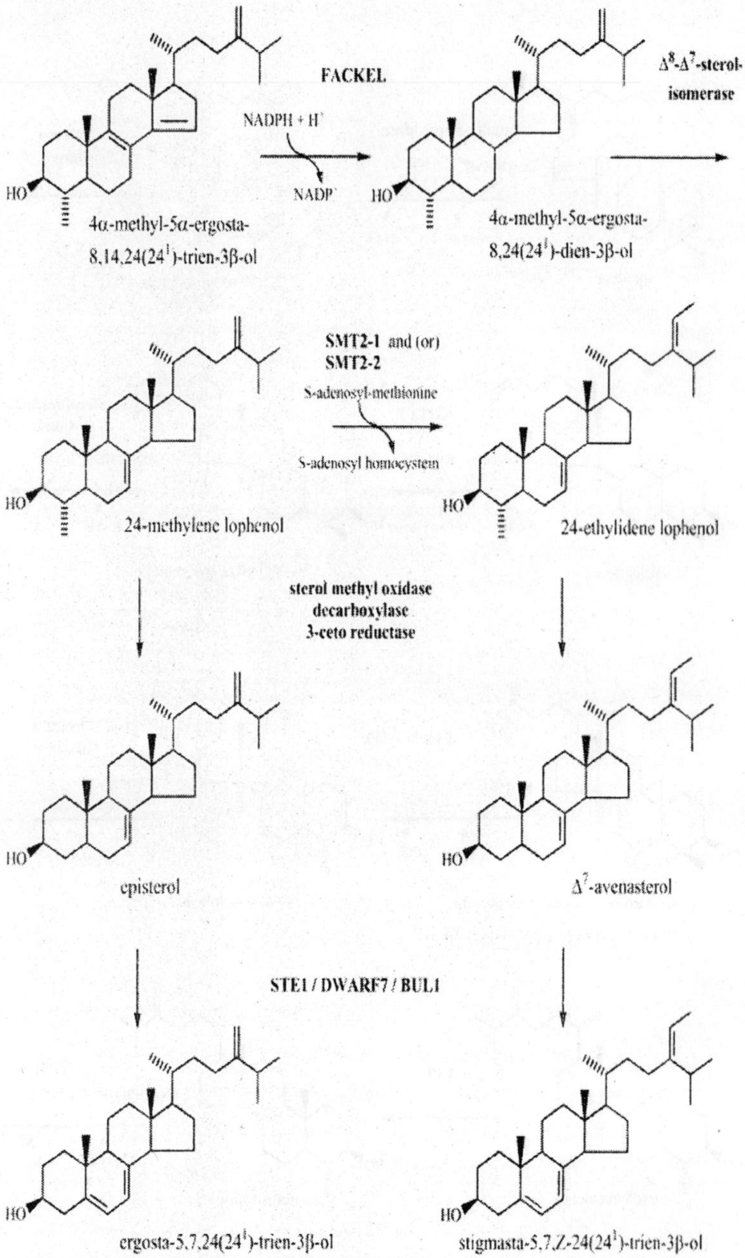
FACKEL
NADPH + H⁺
NADP⁺
Δ⁸-Δ⁷-sterol-isomerase
4α-methyl-5α-ergosta-8,14,24(24¹)-trien-3β-ol
4α-methyl-5α-ergosta-8,24(24¹)-dien-3β-ol
SMT2-1 and (or) SMT2-2
S-adenosyl-methionine
S-adenosyl homocystein
24-methylene lophenol
24-ethylidene lophenol
sterol methyl oxidase
decarboxylase
3-ceto reductase
episterol
Δ⁷-avenasterol
STE1 / DWARF7 / BUL1
ergosta-5,7,24(24¹)-trien-3β-ol
stigmasta-5,7,Z-24(24¹)-trien-3β-ol

Figure 2a

possess only one methyl at C-24. Finally the passage from cycloartenol to end pathway sterols involves three demethylation steps at positions C4 and C14 of the sterol skeleton. The order and the position of these steps strongly differ in non photosynthetic and photosynthetic eukaryotes.

HMGCOA to Squalene

3-Hydroxy-3-methylglutaryl coenzyme A reductase catalyzes the two-step reduction of (S)-3-hydroxy-3-methylglutaryl coenzyme A (S-HMG-COA) into R-mevalonate (Fig 1) and is generally considered to be a regulatory enzyme in sterol biosynthesis.

All HMGR genes from plants isolated so far code for proteins having the same general organization, a highly conserved membrane-spanning domain consisting of two hydrophobic amino-acid sequences that are separated from the globally hydrophilic and highly conserved catalytic domain by a poorly conserved linker region. It appears that plant HMGR isozymes are encoded by a small family of genes and a large body of evidence has been presented that these genes and their products are placed under developmental or environmental control. An Arabidopsis cDNA encoding HMGR has been cloned for the first time by functional complementation of a yeast strain which was defective in mevalonic acid synthesis. Later on Arabidopsis was shown to contain two genes *HMG1* and *HMG2* (At1g76490 and At2g17370) that encode *HMGR*. *HMG1* mRNA is detected in all tissues, whereas the presence of *HMG2* mRNA is restricted to young seedlings, roots and inflorescences. In vitro translation products of these genes were shown to be efficiently inserted into endoplamic reticulum-derived microsomal membranes.

Similar results were obtained with full-length cDNAs (*HMG1* and *HMG2*) from Raphanus sativus, another brassicaceae. Attemps to express a full-length cDNA (*HMG1*) encoding an isozyme of radish HMGR in Escherichia coli resulted in the production of catalytically inactive enzyme, while expression of a truncated form comprising only the soluble domain led to the formation of a highly active enzyme.

Bacterial expression of the catalytical domain of *HMGR* (isoform *HMG1*) from Arabidopsis, and its inactivation by phosphorylation at S577 by Brassica oleracea 3-hydroxy-3-methylglutaryl-CoA reductase kinase were also reported. The use of an alternative promoter in the Arabidopsis *HMG1* gene was shown to generate an mRNA that encodes a novel HMGR isoform (HMG1L) with an extended N-terminal region.

The ER is the only cell compartment for the primary targeting of HMGR in Arabidopsis and that the three HMGR isoforms are cotranslationally inserted into the ER membrane, where they behave as integral membrane proteins. Despite the house keeping role that has been ascribed to *HMG1,* both the quantitative and qualitative features of its expression are modulated in response to a variety of cellular and environmental signals and especially to light. More precisely, light-mediated regulation of HMG1 expression has been shown to be an organ-autonomous response and indicated that the high levels of HMG CoA reductase expression detected in immature leaves may be primarily attributed to the dark-induced expression of *HMG1,* and that *HMG1* is expressed at low levels throughout the plant in response to light.

Finally HMGR is a rate-limiting enzyme for higher plant sterol biosynthesis; its regulation is complex and occurs at transcriptional, translational and post-translational levels. Enzymological studies of the plant enzyme are still needed to better understand features of post-translational regulation. Recent crystal structures of the catalytic portion of human and Pseudomonas mevalonii, HMGRs have given precious clues to better understand some aspects of Arabidopsis HMGR regulation.

Mevalonate kinase (MVK), the enzyme that catalyzes the phosphorylation of *R*-mevalonate by ATP to produce *R*-mevalonate 5-phosphate and ADP, is considered as a potential regulatory enzyme of the isoprenoid biosynthetic pathway. Mevalonate kinase has recently been very much studied in animals because mutations in the gene encoding this enzyme have been shown to be responsible of several pathologies such as the periodic fever syndrome, mevalonic aciduria and hyperimmunoglobulinaemia.

Using an Arabidopsis cDNA library constructed in the yeast expression vector pFL61, a cDNA of 1.64-kb encoding a protein of 378 amino acid residues with significant similarity to yeast and mammalian enzymes, was isolated and characterized by genetic complementation of the *erg12-1* mutation in yeast. Recently molecular cloning and expression analysis of the mevalonate kinase gene from Arabidopsis (AT5g27450) have been reported. The expression pattern of the *MVK* gene suggests that the role of the encoded MVK is the production of a general pool of mevalonate 5-phosphate for the synthesis of different classes of isoprenoids involved in both basic and specialized plant cell functions.

Overexpression of both yeast o. Arabidopsis cDNAs in the yeast had no significant effect on ergosterol accumulation. DNA hybridization blots and Arabidopsis complete genome sequencing show that only one locus exists in the Arabidopsis genome. The single gene from Arabidopsis contains five exons and four introns.

cDNAs or genes encoding 5-phosphomevalonate kinase, the enzyme which catalyzes the phosphorylation of R-mevalonate 5-phosphate to R-mevalonate 5-diphosphate, has still not been isolated and characterized in higher plants. However complete sequencing of Arabidopsis genome has revealed the existence of a gene (At1g31910) whose encoded protein has 30% identity with the yeast PMVK (*ERG8*).

Therefore, with the availability of the *erg8-1* yeast mutant, cloning by complementation seems feasible. Interestingly, so far, two nonorthologous genes encoding PMVK have been described, the *Saccharomyces cerevisiae ERG8* gene and the human PMVK gene. Molecular cloning of human phosphomevalonate kinase allowed the identification of a consensus peroxisomal targeting sequence suggesting that the conversion of mevalonate to farnesyl diphosphate occurs in parallel in the peroxisomes of mammalian cells.

Mevalonate diphosphate decarboxylase (MVDPD, EC 4.1.1.33) catalyzes the ATP-dependent conversion of mevalonate diphosphate into isopentenyl diphosphate (IPP). In *Saccharomyces cerevisiae* this enzyme is encoded by the *ERG*19 gene. Sequence comparison with MVDPD amino acid sequence of *S. cerevisiae* identified an EST clone corresponding to a cDNA that may encode Arabidopsis MVDPD. The predicted amino acid sequence presents about 55% identity with the yeast, and human MVPDs.

When expressed in yeast, the Arabidopsis cDNA complemented an *ERG19* deleted strain. However the wild type sterol content was not fully restored, suggesting that the Arabidopsis MVPD activity may not be optimal in yeast. Complete sequencing of the Arabidopsis genome revealed the existence of two genes (At2g38700 and AT3g54250). Using a two-hybrid assay, it was shown that the *S. cerevisiae* MVDPD forms homodimers *in vivo* and a single substitution impairs dimerization.

The Arabidopsis MVDPD forms also homodimers *in vivo*. Heterodimer formation between the plant and the wild-type yeast enzymes was detected by these assays. Crystal structure would be an attractive goal in order to better understand the intricate mechanism of action of this enzyme.

Isopentenyl diphosphate (IPP) dimethylallyl diphosphate (DMAPP) isomerase catalyzes the conversion of IPP to its electrophilic isomer DMAPP, which is then used to prime the prenyltransferase for chain elongation with IPP. Two Arabidopsis cDNAs (*IDI1* and *IDI2*) encoding isopentenyl diphosphate .isomerase (IDI) were isolated by complementation of an IPP isomerase mutant strain of *S. cerevisiae*. Both cDNAs encode enzymes with an amino terminus that may function as a transit peptide for localization in plastids. Total sequencing of the Arabidopsis genome and DNA blot analysis confirm that *IDI1* and *IDI2* are derived from two genes (AT5g16440, AT3g02780) in Arabidopsis. This seems to be a general feature in most higher plants examined so far and suggests that a single gene would encode plastid and cytosolic IDI in plants. The intracellular localization of isoprenoid synthesis enzymes has been reconsidered in mammalian cells and IDI has been shown to be localized predominantly in peroxisomes and not in the cytosol.

Farnesyl diphosphate synthase catalyzes the condensation of one molecule of dimethylallyldiphosphate (DMAPP) with two molecules of isopentenyl diphosphate (IPP) to give farnesyl diphosphate (FPP). Farnesyl diphosphate is a very important compound in the plant (especially Solanaceae) cell economy because it is located at a branch point that leads to the formation of sterols and sesquiterpenoid phytoalexins.

FPS is encoded by the gene *ERG* 20 in yeast. Disruption of the yeast *ERG* 20 gene is lethal. A cDNA encoding Arabidopsis FPS1 was subsequently isolated by complementation of the *erg* 20 deletion yeast mutant with the cDNA expression library cloned into pFL61. This cDNA encodes a polypeptide of 343 amino acids residues presenting about 50% identity with yeast, rat and human FPSs. A second cDNA corresponding to the *FPS*2 was isolated by RT-PCR, it codes for a protein having 342 amino acid residues. *FPS*1 and *FPS*2 share an overall amino acid identity of 90.6%. The *FPS*1 gene (AT5g47770) consists of 12 exons and 11 introns whereas the *FPS*2 gene (AT4g17190) consists of 11 exons and 10 introns. In both genes, introns are located at equivalent positions relative to the coding sequences. *FPS*1 mRNA accumulates preferentially in roots and inflorescences whereas *FPS*2 mRNA is predominantly expressed in inflorescences. Later on it has been shown that the *FPS*1 gene generates a novel mRNA that encodes a mitochondrial farnesyl diphosphate synthase isoform (FPS1L).

The FPS1L mRNA accumulated preferentially in inflorescences whereas the previously reported FPS1 mRNA (FPS1S mRNA) is predominantly expressed in roots and inflorescences. Translation of the FPS1L mRNA from the upstream ATG start codon generates a novel FPS1 isoform (FPS1L) with an NH2-terminal extension of 41 amino-acid residues, which has all the characteristics of a mitochondrial transit peptide.

The functionality of this extension as a mitochondrial transit peptide was demonstrated by *in vitro* import experiments using purified plant mitochondria. Spatial and temporal patterns of the β-glucuronidase (GUS) reporter gene expression directed by 5' regions of the Arabidopsis *FPS1* and *FPS2* genes have been studied.

The highest levels of FPS2: GUS activity were detected in flowers, especially in pollen grains. Overall, the pattern of expression of *FPS2 :GUS* suggests a role for *FPS2* in the synthesis of particular isoprenoids with specialized functions. In contrast, the *FPS1S :GUS* gene is widely expressed in all plant tissues throughout development, thus supporting a role for *FPS1S* in the synthesis of isoprenoids (for instance sterols) serving basic plant cell functions.

Recently cDNA sequences encoding FPS have been cloned from several plant species such as *Lupinus albus*, *Zea mays*, *Artemisia annua*, *Hevea brasiliensis*, and *Parthenium argentatum*. According to these studies, it seems that plant FPS would be encoded by multigene families like other key enzymes of the isoprenoid biosynthesis such as HMGR, IDI and several others (*vide infra*). The biological significance of the occurrence of similar isoforms in plants will be discussed in a conclusive paragraph.

Nevertheless, FPSs are considered to play a key role in isoprenoid biosynthesis. It has been shown that in mammals FPS is a highly regulated enzyme involved in the control of the sterol biosynthetic pathway.

Squalene synthase catalyzes the reductive condensation of two molecules of FPP to squalene in the presence of NADPH and Mg^{2+}, via the intermediate presqualene diphosphate (Fig. 1). In contrast to preceding enzymes, SQS catalyzes the first committed step in sterol formation and for this reason may be highly regulated.

A SQS probe from *Mus musculus* was successfully used to isolate two overlapping Arabidopsis cDNA clones coding for a protein of 410 amino acids with about 40% identity with mammalian and yeast squalene synthetases. The *Arabidopsis* SQS enzyme was expressed in *Escherichia coli*. A cell-free extract of *E. coli* transformed with a recombinant plasmid containing the *Arabidopsis* SQS cDNA efficiently converted [^{14}C]farnesyl diphosphate into squalene in the presence of NADPH and Mg^{2+}.

Both the structures and reaction mechanisms of squalene synthases were markedly conserved between plants and mammals. Later on it was shown that *Arabidopsis* SQS is encoded by a small gene family of two genes: SQS1 (AT4g34640) and SQS2 (AT4g34650), which are organized in a tandem array. SQS1 and SQS2 have an identical organization regarding intron positions and exon sizes and encode SQS isoforms showing a high level of sequence conservation (79% identity). Remarkably the SQS1 isoform is unable to complement the SQS-defective (gene disrupted) *Saccharomyces cerevisiae* strain 5302, although SQS activity was detected in the microsomal fraction of the transformed yeast strain.

However a chimeric SQS resulting from the replacement of the 66 C-terminal residues of the *Arabidopsis* enzyme by the 111 C-terminal residues of the *Schizosaccharomyces pombe* enzyme was able to confer ergosterol prototrophy to strain 5302. This and other expriments indicated that the C-terminal region of the enzyme is involved in the channeling of squalene through the yeast sterol pathway.

Squalene synthase cDNAs were isolated also in *Glycyrrhiza glabra, Nicotiana tabacum, Nicotiana benthamiana, Orizum sativa, Zea may, Solanum tuberosum* and *Glycine max*. Treatment of tobacco cell suspensions cultures with a fungal elicitor dramatically reduced SQS enzymatic activity. Analysis of SQS mRNA levels in elicitor-treated cells indicated that the suppression of SQS enzyme activity may result from a post-transcriptional modification of SQS.

Squalestatin S1, a potent inhibitor of SQS in mammalian cells, has been shown to strongly inhibit sterol biosynthesis in tobacco BY-2 cells. Conversely enzymatic activity and mRNA levels of HMGR were several-fold increased. Finally a crystal structure of human squalene synthase has been recently reported and should be useful in the design of new inhibitors of potential therapeutic importance (Pandit et al., 2000).

Squalene to Sterols

Squalene epoxidase catalyzes the conversion of squalene to 2,3(S)-oxidosqualene (OS) (Fig 2) which is the first oxygenation step in the sterol biosynthetic pathway. This microsomal-bound enzyme has been thoroughly studied in animal cells. Rat squalene epoxidase requires FAD, NADPH-cytochrome P450 reductase, NADPH and a supernatant protein factor (SPF). Rat squalene epoxidase cDNA was isolated by selecting *S. cerevisiae* transformants expressing rat cDNAs in the presence of terbinafin, a potent fungicide and inhibitor of fungal squalene epoxidase.

Rat squalene epoxidase, as deduced from the nucleotide sequence, contains 573 amino acids. The amino acid sequence reveals one potential transmembrane domain and a FAD binding domain, suggesting that the squalene epoxidase is a flavo-protein. This deduced sequence is 30% identical to *ERG1*, the gene encoding squalene epoxidase in *S. cerevisiae*. Rat squalene epoxidase expressed in *E. coli* was shown to catalyze the *in vitro* conversion of squalene to 2,3(S)-oxidosqualene, thus confirming the polypeptide as a functional squalene epoxidase.

The gene encoding the supernatant protein factor (SPF) that promotes the squalene epoxidation catalyzed by rat liver microsomes has been recently cloned. The encoded protein of 403 amino acids belongs to a family of cytosolic lipid-binding/transfer proteins. Only a few studies concerning squalene epoxidase have been reported in plants. The sequencing of the Arabidopsis genome has revealed 6 genomic clones (AT1g58440, AT2g22830, AT4g37760, AT5g24140, AT5g24150 and AT5g24160) whose deduced polypeptidic sequences possess significant identities with the rat liver enzyme. The amino acid sequences reveal one potential transmembrane region and a FAD binding domain suggesting the existence in Arabidopsis of a family of genes encoding squalene epoxidase. However no data dealing with the function of theses genes has been reported so far.

Sequences of 3 Arabidopsis and 2 *Brassica napus* cDNAs encoding squalene epoxidase homologues have been reported. Comparison of cDNA and genomic sequences indicate that the 3' splice site of an intron in these genes has undergone junctional sliding, a phenomenon having interesting evolutionary significance.

2,3(S)-oxidosqualene-cycloartenol cyclases (cycloartenol synthase). The 2,3(S)-oxidosqualene (OS) cyclases compose a family of biocatalysts that convert (S)-2,3-oxidosqualene (OS) to polycyclic triterpenes. The mammalian (Abe and Prestwich, 1995) and fungal OS cyclase produces lanosterol, precursor of cholesterol in vertebrates and of ergosterol in most fungi.

The higher plant OS cyclase catalyzes the formation of cycloartenol, precursor of phytosterols. This latter enzyme must break 11 bonds and form 11 new ones to form cycloartenol, which is a pentacyclic cyclopropyl triterpene containing 9 chiral centers. In addition to cycloartenol synthase, higher plants contain other OS-cyclases converting OS into a vast array of pentacyclic triterpenoids (oleanane, ursane, lupane, baccharane, friedelane derivatives).

The stereoelectronic factors that understate the catalytic mechanism. The formation of lanosterol and cycloartenol is initiated in the chair-boat-chair-like conformation of OS whereas the formation of amyrins and related pentacyclic triterpenes is initiated in the chair-chair-chair-like conformation of OS. In all cases the proton-initiated cyclization is postulated to proceed through the following events :

i) a series of antiperiplanar 1,2 additions ; 1,2 rearrangements and 1,2 eliminations;

ii) the formation of conformationally rigid and partially cyclized carbocationic intermediates;

iii) stabilization of the developing carbocationic centers on the substrate by negative point charges at the active site of the enzyme.

These cyclization intermediates could be stabilized by cation-interactions. The fact that OS mimics possessing positively charged nitrogen atoms in place of potential carbocation behave as potent inhibitors. Most prokaryotic organisms do not synthesize sterols, but many of them synthesize hopanoids that are derived from pentacyclic triterpenic hydrocarbons which originate from squalene cyclisation owing to squalene-hopene cyclases.

A 2,3-OS-lanosterol cyclase (lanosterol synthase) gene (ERG7) from *S. cerevisiae* was isolated and encoded a 83-kDa protein. Later on a splendid approach led to the isolation of an Arabidopsis cDNA (ATCYC) encoding a cycloartenol synthase. This was achieved by functional expression in a yeast mutant lacking lanosterol synthase (GL7) by the use of a chromatographic screen.

This cDNA contained a 2277-bp open reading frame encoding a 86-kD protein which was about 35% identical with *S. cerevisiae* lanosterol synthase and remarkably 46% identity with a *Rattus norvegicus* lanosterol synthase showing that the plant cycloartenol synthase has more identity with a mammalian than a fungal lanosterol synthase. A single mutation (adenine1362guanine) occurring on the Arabidopsis cycloartenol synthase gene has been shown to confer to cycloartenol synthase the ability to form 25% lanosterol and 21% parkeol in addition to cycloartenol. This mutation leads to a I454V change in the peptidic sequence. A yeast mutant (*erg7*) defective in lanosterol synthase was transformed with a plasmid harbouring the cycloartenol synthase cDNA.

The transformed yeast was still auxotrophic to ergosterol because cycloartenol is not usable by yeast to make sterols. *Erg7* did not need any more ergosterol if transformed with a cycloartenol synthase cDNA possessing the A1362G mutation. Likewise, a tyrosine-to-threonine mutation was shown to convert the Arabidopsis cycloartenol synthase synthase to an OS cyclase that forms lanosterol as its major product.

Such experiments have far-reaching evolutionary implications. The corresponding single gene (At2g07050) consists of 18 exons and 17 introns. The Arabidopsis genome sequencing program reveals the presence of a gene (AT3g45130) whose deduced polypeptide sequence possessed 62% identity with the cycloartenol synthase protein and appeared closest to cycloartenol synthase in a phylogenetic tree containing many of the triterpene synthases already known (Fig. 3). The function of the protein encoded by this gene is still unknown.

In recent years a breakthrough has been achieved in the field of triterpene synthases. A cDNA encoding ß-amyrin synthase (*PGAM*) has been characterized in *Panax ginseng*. Expression of the cDNA in yeast led to formation of a polypeptide capable of converting OS into β-amyrin. A cDNA encoding a lupeol synthase (*ATLUP1*) has been identified in Arabidopsis after expression in *S. cerevisiae*. The protein (ATLUP1) encoded by this cDNA was shown to convert OS into lupeol, minor triter-penoid compounds were also identified. Another cDNA (*ATLUP2*) was subsequently cloned from Arabidopsis and expressed in *S. cerevisiae*. The transformed *S. cerevisiae* was shown to cyclize OS into a-, β-amyrins and lupeol. Therefore the polypeptide (ATLUP2) encoded by this gene is a novel multifunctional triterpene synthase. Shuffling domain mutagenesis has been done on ATLUP1 and PGAM in order to investigate the region important for protein specificity. The second quarter from the N-terminus of the protein was shown to have a crucial importance in determining the enzymatic process toward lupeol or amyrin.

The genome sequencing project has revealed a rather complex situation. *ATLUP1* and *ATLUP2* are representatives of a small family of five genes (*ATLUP1-5*), in addition seven genes (*ATPEN1-7*) whose deduced polypeptide sequences have about 55% identity with the *ATLUP* family and which could encode triterpene synthases have been also identified.

Each genomic sequence is labeled with a code and is also characterized by the number of exons and introns. In order to obtain information on evolutionary relationships, a phylogenetic tree was constructed using a heuristic search (Fig. 3). In this tree the prokaryotic hopane synthases from *Alicyclobacillus* (*AAHOP*) was considered as reference taxa. According to this tree, the seven

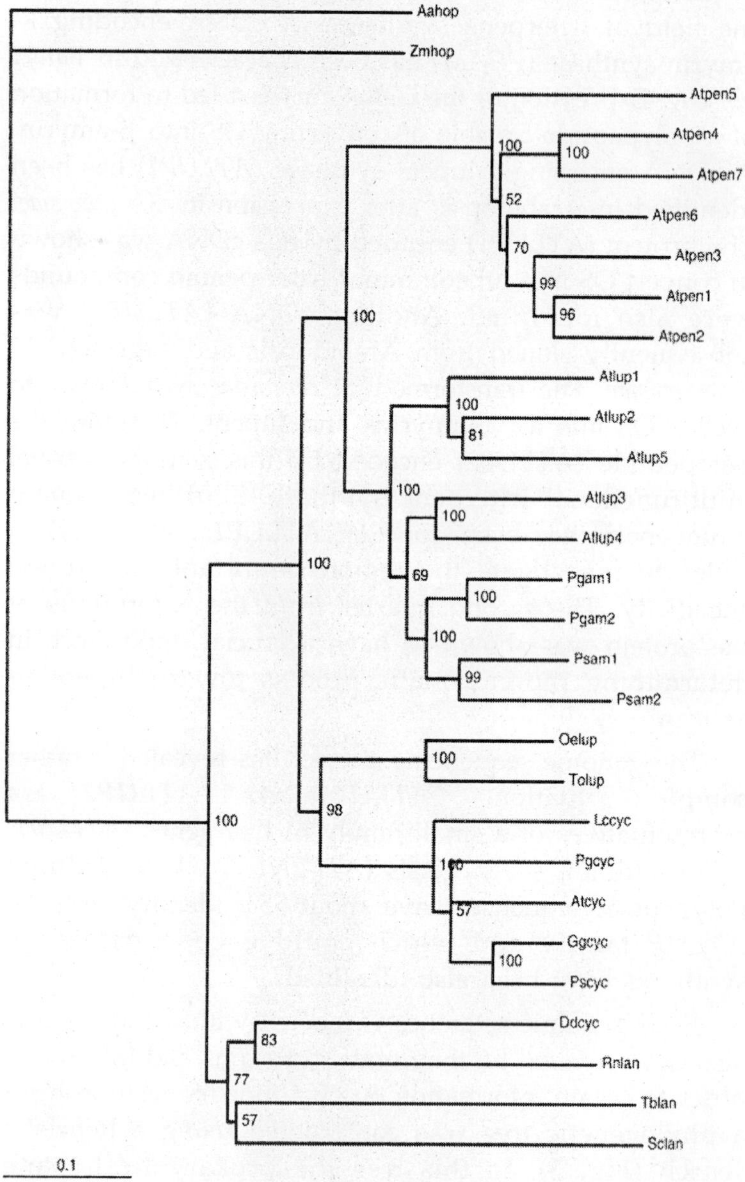

Figure 3. Phylogenetic tree constructed for several plant, animal, fungal and bacterial triterpene synthases.

pentacyclic triterpene synthases from Arabidopsis (*ATPEN1*to *7*) form a group of genes clearly distinct from the five lupeol synthases (*ATLUP1* to *5*).

Interestingly, the *ATLUP* subfamily also encompasses the two β-amyrin synthases of *Panax ginseng* (*PGAM*) and of *Pisum sativum* (PSAM) and is clearly distinct from the lupeol synthases from *Olea europaea* (*OELUP*) and *Taraxacum officinalis* (*TOLUP*).

S-Adenosylmethionine-sterol-C-methyltransferases. Sterols from fungi and higher plants differ from vertebrate sterols by the presence of an extra alkylgroup at C-24. Whereas most fungi sterols possess a methyl group at C-24, higher plants contain both 24-methyl- and 24-ethyl sterols. This alkylation of the side chain is catalyzed by S-adenosylmethionine (AdoMet)-sterol-C-methyltransferases (SMTs).

In *S. cerevisiae,* the SMT converts zymosterol into fecosterol. In higher plants (and therefore in Arabidopsis), the presence of 24-ethylsterols results from two distinct methyl-transfers from AdoMet. According to the chemical structures of intermediates of plant sterol biosynthesis and substrate specificity studies, it is generally assumed that cycloartenol (Fig 2) is the substrate of the first methylation reaction, resulting in 24-methylene cycloartanol, whereas 24-methylene lophenol is the preferred substrate for the second methylation, yielding 24-ethylidene lophenol.

Because the chemical structures of 24-methylene cyclartanol and 24-methylene lophenol are very different, it has been suggested that the two methylation ractions would be catalyzed by two different enzymes. However, since no plant SMT has been purified so far, the hypothesis of an unique plant SMT catalyzing both alkylations should also be considered. The first SMT gene (*ERG6*) cloned was from *S. cerevisiae.*

Later on complete identification was performed by enzymatic assays on the protein derived from *ERG6* after expression in *E. coli*. The best substrate of the recombinant enzyme was zymosterol which was methylated to give fecosterol in agreement with previous enzymatic assays performed on microsomes from WT *S. cerevisiae*. A *SMT* cDNA of 1411 bp has been cloned from an Arabidopsis cDNA library, the deduced amino acid sequence of this cDNA showed three conserved regions found in AdoMet-dependent methyltransferases and 38% identity with *ERG6*.

This cDNA was used to transform a wild-type *S. cerevisiae* as well as the yeast null mutant *erg6*, which is deficient in the SMT-zymosterol-C-24 methyltransferase. In both cases, several 24-ethyl and 24-ethylidene sterols were synthetized, indicating that the Arabidopsis cDNA encodes a plant SMT capable of performing two sequencial methylations at C24 and C24-1 of the yeast sterols. Microsomes from *erg6* expressing the Arabidopsis SMT were shown to possess AdoMet-dependent sterol-C-methyltransferase activity.

Delipidated preparations of these microsomes converted cycloartenol into 24-methylene cycloartanol and 24-methylene lophenol into 24-ethylidene lophenol. The catalytic efficiency of the expressed SMT was 17-times higher with 24-methylene lophenol than with cycloartenol. This result provides evidence that the Arabidopsis cDNA encodes a 24-methylene-lophenol-C-241-methyltransferase catalyzing the second methylation step of plant sterol biosynthesis.

This cDNA was named *ATSMT2-1*. A second cDNA (*ATSMT2-2*) of 1249 bp was cloned and was shown to encode a protein of 359 amino acids, 82% identical with the protein deduced from *ATSMT2-1*. Meanwhile two cDNAs from *Nicotiana tabacum* (*NTSMT2-1* and *NTSMT2-*

2) and one from *Oriza sativa* (*OSSMT2*) were characterized and were shown to be about 80% identical to the Arabidopsis cDNAs. The cDNA *NTSMT2-1* was expressed in *erg6* and the corresponding protein was shown to convert 24-methylene lophenol into 24-ethylidene lophenol as efficiently as ATSMT2-1. Meanwhile cDNAs from *Glycine max, Ricinus communis, Zea mays, N. tabacum* and *O. sativa* and Arabidopsis were isolated and characterized. The proteins encoded by these cDNAs are about 80% identical in all possible combinations, but have only 40% identity with SMT2 and 48% identity with ERG6.

They constitute therefore a group of cDNAs (*SMT1*) distinct of the group formed by *SMT2*. *SMT1* from *Z. mays*, from Arabidopsis and from *N. tabacum* were shown to partially restore the abillity of *erg6* to make ergosterol. The *G.max* SMT1 was expressed in *E. coli* and the recombinant protein was shown to alkylate lanosterol (an unnatural substrate in plants) to give eburicol.

In the presence of AdoMet, delipidated microsomes from *erg6* transformed with tobacco *SMT1* efficiently converted cycloartenol into 24-methylene cycloartanol, but did not produce any 24-ethylidene lophenol upon incubation with 24-methylene lophenol. This demonstrated that cDNA *NTSMT1* (and most probably the other plant *SMT* of the same group such as *ATSMT1*) encoded a cycloartenol-C24 methyltransferase.

Sequence alignment of SMT1 and SMT2 reveal three highly conserved domains. One of them (LDXGCGXGGPXRXI) corresponds to the G-rich consensus motif described for all AdoMet-dependent methyltransferases. A second (IEATCHAP) and a third (YEW/F/YGWGXSFHF) motives are believed to be typical of methyltransferases acting on a sterol substrate.

Whereas SMT2 possess a hydrophobic domain of approximately 25 amino acids at the N-terminal position, SMT1 are devoid of such a hydrophobic domain . As confirmed by the *Arabidopsis* sequencing project and the above results, the Arabidopsis genome contains three distinct genes (At5g13710, At1g20330, At1g76090) encoding sterol-C24 methyltransferases(ATSMT1, ATSMT21, ATSMT22 respectively). *smt1* mutations, which lack SMT1, were isolated from a transposon Ac transgenic line.

The *smt1* plants have pleiotropic defects: poor growth and fertility, root sensitivity to Ca^{2+} ions, and a loss of proper embryo morphogenesis. *smt1* was shown to have an altered sterol content. It accumulates cholesterol, has much less 24-ethyl sterols, but its campesterol content is similar to this in control plants, reflecting that methylation of endogenous sterols is still active even in the absence of any functional SMT1 present.

Actually, if SMT1 were the only enzyme responsible *in planta* for the first methylation reaction of cycloartenol to produce 24-methylene cycloartanol, *smt1* should be completely deficient for C-24 alkylation. Because the majority of sterols in *smt1* plants are alkylated, SMT1 cannot be the only enzyme *in* Arabidopsis capable of the first C-24 methylation step.

One explanation for the presence of 24-methylated sterols in *smt1* is that in the absence of *SMT1*, these methylations would be performed by SMT21 or SMT22 (that do not ordinarily perform this reaction in the wild type because of their preference for 24-methylene lophenol) even though at a much slower rate. The expression of *ATSMT21* was modulated in *35S : :SMT21* Arabidopsis in order to study its physiological function.

Plants overexpressing the transgene accumulated sitosterol at the expense of campesterol. These plants displayed a reduced stature and growth that could be restored by brassinosteroid treatment. Plants showing co-suppression of *SMT21* were characterized by a high campesterol content and a depletion in sitosterol. Pleiotropic effects on development such as reduced growth, increased branching, and low fertility of high campesterol plants were not modified by exogenous brassinosteroids, indicative of specific sterol requirements to promote normal development.

Thus *ATSMT21* has a crucial role in balancing the ratio of campesterol to sitosterol in order to satisfy both growth requirements and membrane integrity.

4,4-dimethyl sterol and 4a-methyl sterol 4-demethylation. The passage of 24-methylene cyclartanol to end-pathway sterols involves removal of two methyls at position 4 and one methyl at position 14. The steps involved in these operations are depicted in Fig.2. The enzymatic system operating in the removal of the two methyls at C-4 in plants presents profound differences with the fungal or mammalian systems.

In contrast to animals and fungi where the two C-4-methyl group are sequentially removed, a series of results have unequivocally established that two distinct oxidative systems are involved in the removal of the first and the second C4-methyl of phytosterol precursors. The first will act on 24-methylene cycloartanol whereas the second will operate on 24-methylene and 24-ethylidene lophenol (Fig 2).

In the case of the first enzymatic system it has been shown that the oxidative conversion of 24-methylene cycloartanol to cycloeucalenol exhibits similar cofactor requirements, and inhibitor sensitivity as do corresponding animal systems. The demethylation is specific for the C-4a methyl group and is initiated by the C-4 methyl oxidase, which converts the methyl group to the alcohol, the aldehyde and finally to the carboxylic acid. This sequential oxidation was shown to require NADH, oxygen and cytochrome b5 as an electron carrier between the terminal oxidase and NADH. It is not inhibited by CO but is susceptible to cyanide, indicating that a cytochrome P-450 is not involved. In the next reaction, the carboxyl group is removed by a second enzyme, the C4 decarboxylase acting on 4α-carboxy-4β,14α-dimethyl-9β,19-cyclo-5α–ergost-24(24^1)-en-3β–ol and resulting in formation of cycloeucalenone possessing a keto group at C-3. A third enzyme, 3-keto reductase, then reduces the keto group of cycloeucalenone to give cycloeucalenol.

This process is repeated with the second methyl group but the substrate is then 24-methylene or 24-ethylidene lophenol and the product is then episterol or Δ^7-avenasterol, respectively. Cloning and characterization of *ERG25*, the *S. cerevisiae* gene encoding C4-methyl oxidase has been reported. This gene encodes a 309-amino acid polypeptide showing a C-terminal endoplasmic reticulum (ER) retrieval signal KKXX and three histidine-rich clusters found in eukaryotic membrane bound fatty acid desaturases.

A human homologue of *ERG25* was cloned and sequenced. It encodes a polypeptide of 293 amino acids with an identity of 38% to *ERG25* and contains histidine clusters. Very recently the functional identification of sterol-4α-methyl oxidase cDNAs from *A. thaliana* was achieved by complementation of a yeast *erg25* mutant lacking sterol-4a-methyl oxidation.

Cycloeucalenol obtusifoliol isomerase. One of the most striking event occurring during sterol biosynthesis is the opening of the cyclopropane ring of cycleucalenol to give obtusifoliol, a step that is restricted to the plant kingdom and catalyzed by an enzyme, the cycloeucalenol-obtusifoliol isomerase or cyclopropyl sterol isomerase (CPI), the catalytic mechanism of which has been thoroughly studied (Fig. 2). Photosynthetic eukaryotes and certain non photosynthetic protists such as *Acantamoeba polyphaga* or *Dictyostelium discoidum* use CPI to convert pentacyclic cyclo-propyl sterols to tetracyclic end-pathway sterols. Arabidopsis CPI was cloned by functional complementation of a *S. cerevisiae* mutant.

Expression of an Arabidopsis cycloartenol synthase cDNA in a *S. cerevisiae* lanosterol mutant (*erg7*) provided a sterol auxotroph because cycloartenol is not usable by yeast to make ergosterol. This yeast strain was transformed with an Arabidopsis expression library constructed in a yeast vector (pFL61) and sterol prototrophs were selected. A strain accumulating biosynthetic ergosterol was obtained. The novel phenotype was conferred by an Arabidopsis cDNA (CPI) that encoded a 36-kDa protein of 280 amino acids.

It was assumed that the *S. cerevisiae* sterol C4 demethylase complex (Erg25p, Erg26p, and Erg27p) that normally metabolize lanosterol would have sufficiently broad substrate specificity to accept its isomer cycloartenol. Thus cycloartenol would be demethylated at C4 (at least partially) to give small amounts of 31-nor-cycloartenol which was shown to be readily transformed into 31-nor lanosterol, which could then enter the normal yeast sterol pathway. CPI is encoded by a unique gene (At 5g50375) possessing 8 exons.

Obtusifoliol-14a-demethylase. In animals and fungi, the 14a–methyl group is the first of the C14 and C4 methyls to be removed; however in higher plants, the 14a–methyl is removed after one C-4 methyl has disappeared (Fig 2). The mechanism of the 14a–methyl removal involves two oxidation steps leading to an alcohol, then an aldehyde at C-29 and a further oxidative step involving a deformylation leading to formic acid and the sterol product with a typical 8,14-diene. The three steps are catalyzed by a single enzyme, which has been shown to be a cytochrome P-450. Lanosterol is the substrate of this enzyme in *S. cerevisiae* and in mammals, eburicol would be the preferred substrate in filamentous fungi, while obtusifoliol is the substrate in higher plants.

Much attention has been payed to the 14α–demethylase since the enzyme is the target of a vast array of compounds having antimycotic effect and used in agriculture and medicine. These compounds, which are derivatives of pyridine, pyrimidine, imidazole, triazole...interact by their nitrogen sp2 doublet with the protohemic iron of cytochrome P-450 and hamper the access to oxygen.

The plant pathway differs from fungal and animal pathways downstream of 2(3)-oxidosqualene. No lanosterol is present in plants to serve as substrate for the 14-demethylation step. It is obtusifoliol that is the substrate for 14-demethylation. A thorough enzymological study has defined the stuctural requirements of this enzyme, which was shown to be remarkably specific for the structure of obtusifoliol and which does not use lanosterol.

A panoply of azole derivatives were shown to inhibit this enzyme; some of them, developed as herbicides, were shown to be specific to obtusifoliol-14-demethylase (OBT14DM). Tobacco calli resistant to phytotoxic azole derivatives have been isolated in mutagenesis experiments. The mechanism of resistance consists of an overproduction of sterols, which is not due to OBT14DM overproduction or modification but to an increase of

HMGR enzymatic activity. cDNAs encoding lanosterol, eburicol and obtusifoliol 14-demethylases have been isolated from mammals, fungi and plants respectively. They shared a remarkable amino acid identity ranging from 38% to 65% though they belong to widely diverging species. Therefore they were classified in the same family, which is CYP51.

A functional *Sorghum bicolor* OBT14DM was cloned and expressed at high levels in *E. coli.* The recombinant enzyme was shown to catalyze efficiently the 14a–demethylation of obtusifoliol. Likewise a *Triticum vulgare* OBT14DM was expressed in a *S. cerevisiae* mutant (*erg11*) defective in lanosterol 14-demethylase. Appropriate engineering of the *TvOBT14DM* cDNA and of the recipient *erg11* mutant allowed to optimize expression.

Experimental structural informations on the active sites of the fungal, plant, and mammalian CYP51 would greatly facilitate developing more efficient antifungal drugs. However all forms of CYP51 were membrane-bound microsomal enzymes which complicates structural studies of this protein by X-ray crystallography. A soluble CYP51 ortholog has been discovered recently in *Mycobacterium tuberculosis*. It has been shown to demethylate lanosterol and obtusifoliol and to be inhibited by azole antifungals.

E. coli-expressed MTCYP51 has been recently crystallized in the presence of fluconazole and a crystal structure at 2.2 A has been reported. This new structure provides a basis for rational design of more efficacious antifungal agents or new herbicides as well as insight into the molecular mechanism of P450 catalysis. Arabidopsis genome sequencing has revealed the existence of two genes (At2g17330, At1g11680) encoding proteins having 65–75 % identity with an already characterized OBT14DM from *Sorghum bicolor*. Both genes possess two exons and an intron. The cDNA (ATCYP51) corresponding to At1g11680 has been cloned and has been shown to be able to sustain growth of the *erg11* mutant suggesting ATCYP51 to be a functional sterol 14a–demethylase. Arabidopsis plants were transformed with AtCYP51 in antisense orientation. The resulting transgenic plants showed a semi-dwarf phenotype in the early growth stage. Their obtusifoliol content increased while campesterol and campestanol levels were unchanged. As their 24-ethyl sterols content was not reported, it is still not possible to draw conclusions as to the link between growth reduction and sterol biosynthesis inhibition in these transgenic plants.

Δ^8,14-sterol-Δ^{14}-reductase. 4α,14α-dimethyl-5α-ergosta-8,14,24(24^1)-trien-3β–ol is the product of the C14 demethylation of obtusifoliol in plants. The following step is the hydrogenation by NADPH of the Δ^{14} double bond to give fecosterol (4α–methyl-5α–ergosta-8,24(24^1)-dien-3β–ol) (Fig. 2). The product of the C14 demethylase reaction in *S. cerevisiae*, 4,4-dimethyl-cholesta-8,14,24-trienol, is reduced at the C-14 position to form 4,4-dimethylcholesta-8,24-dienol by the action of the product of the *ERG24* gene.

ERG24 encoded a protein of 438 amino acids. The C14 reductase is inhibited by Fenpropidine and Fenpropimorph, two fungicides used in agriculture. By chromosomal gene disruption a *S. cerevisiae* defective in 14-reductase has been constructed. This strain was shown to grow in aerobiosis if it beared an additional mutation allowing sterol uptake. In this last growth condition, the cells required a sparking ergosterol supplementation and accumulated 5α-ergosta-8,14-dienol as the end-product of the sterol pathway. Most interestingly it has been shown that the human lamin B receptor exhibited sterol C-14-

reductase activity in *S. cerevisiae.* In fact, the sterol C14 reduction step and ergosterol prototrophy were restored in lamin B receptor-producing *erg24* transformants which lack endogenous C14-reductase.

Rubus fruticosus suspension cell cultures were highly susceptible to A25822B, an azasterol antimycotic agent and accumulated 5α–stigmasta-8,14-dien-3β–ol and 5α–stigmasta-8,14, Z-24(24[1])-trien-3–ol at the expense of Δ^5-end pathway sterols. An enzymatic assay for 8,14-sterol 14-reductase was devised allowing to characterize this enzyme and to study its inhibition by analogues of a presumptive carbo-cationic intermediate of the reduction reaction.

Cloning of the gene (*FACKEL*) encoding this enzyme was performed recently through the isolation of dwarf mutants (EMS and TDNA insertion mutants). *FACK-EL* (At 3g52940) was isolated and was shown to encode a predicted integral membrane protein of 369 amino acids with eight to nine transmembrane segments related to the vertebrate lamin receptor and several sterol C-14 reductases including *S. cerevisiae* sterol C-14 reductase.

The Arabidopsis protein possessed a signature motif (LLXSGYWGXXRH) of sterol reductases. Functional evidence that *FACKEL* encoded a sterol C-14 reductase was provided by complementation of *erg24*. GC/MS analysis confirmed that *fk* mutations lead to accumulation of Δ^8,14-sterol intermediates in the biosynthetic pathway preceding the C-14 reductase step. The sterol profile of *fk* calli was similar to that of plant cells treated with A25822B.

The *fk* mutation resulted also in reduction of brassinosteroid (BRs) content. However, unlike other BR-deficient mutants, the defect of hypotyl elongation could not be overcome by exogenous BRs. Thorough microscopical and cytological observations indicated that mutations in the *FACKEL* gene affect body organization of the Arabidopsis seedlings and that *FACKEL* was required for cell division and expansion and was involved in proper organization of the embryo. These results indicated a novel role for sterols in the embryogenesis of plants.

Δ^8-Δ^7-sterol isomerase. When the 14α–methyl group is removed and the 14 double bond is reduced, the resulting Δ^8-sterols are isomerized to Δ^7-sterols in mammals, fungi, and higher plants. This process is catalyzed by a Δ^8-Δ^7-isomerase. The reaction involves two steps : first a protonation of the 8 double bond leading to an intermediate bearing a carbocation at C8 then the elimination of a proton at C7 leads to the Δ^7-sterol.

From enzymatic assays and biogenetic considerations it appears that zymosterol, fecosterol and 4α–methyl-5α–stigmasta-8, Z-24(24^1)-dien-3β–ol are the substrates of this enzyme in vertebrates, *S. cerevisiae* and higher plants respectively. A mutant (*erg2*) defective in Δ^8-Δ7-sterol isomerase was described for *S. cerevisiae*. The *erg2* mutation leads to the accumulation of sterols containing the Δ^8-double bond. *ERG2*, the gene encoding the Δ^8-Δ^7-sterol isomerase (SI) was cloned by functional complementation of *erg2*. Later, SI-encoding genes were isolated from the rice blast fungus *Magnaporthe grisea* and the maize pathogen *Ustilago maydis*.

Genetic and biochemical evidence showed that *ERG2* would encode an enzyme catalyzing conversion of Δ^8-sterols to Δ^7-sterols. Recently a *S. cerevisiae* mutant defective in the internalization step of endocytosis was shown to contain defects in *ERG2* leading to a lack of ?8-Δ^7-isomerase activity. A murine SI encoding cDNA has been cloned by functional complementation of the corresponding deficiency (*erg2*) in *S. cerevisiae*. The amino acid sequence deduced from the cDNA open reading frame was shown to be highly similar to human emopamil-binding protein (EBP), a protein which is the target for neuroprotective drugs.

A yeast strain in which the SI coding sequence has been replaced by that of human EBP or its murine homologue, recovered the ability to convert Δ^8-sterols to Δ^7-sterols both *in vivo* and *in vitro*. Interestingly, the amino acid sequence deduced from the murine (SI) or human (EBP) failed to reveal any striking similarity with *S. cerevisiae* Erg2p (13% identity). Mutations in the gene encoding *Homo sapiens* SI were shown to cause X-linked dominant Conradi-Hunermann syndrome.

Alanine scanning mutagenesis was used to identify residues in the four putative transmembrane a-helices of human EBP that are required for Δ^8-Δ^7-sterol isomerase catalytical activity and binding of inhibitors. Out of 64 ala-nine mutants of EBP mutants H77A, E81A, T126A, N194A, W197A, contained less than 10% end-pathway 5,7-sterols. Therefore amino acids H77, E81, T126, N194 and W197 should play an important role in sterol Δ^8-Δ^7 isomerization.

An Arabidopsis Δ^8-Δ^7-sterol isomerase cDNA has been isolated by functional complementation of the corresponding *S. cerevisiae* sterol mutant (*erg2*). The full length Arabidopsis cDNA that complement the *erg2* mutation contained an open reading frame encoding a protein of 223 amino acids sharing 35% amino acid identity to the *Mus musculus* SI and the *H. sapiens* EBP and very low identity (less than 15%) with Erg2p. The sigma ligands, haloperidol, ifenprodil and verapamil inhibited the production of ergosterol in wild-type *S. cerevisiae* and in the *erg2* mutant complemented with the Arabidopsis SI. The Arabidopsis protein contained three transmembrane segments and presented several structural and biochemical similarities with the mammalian EBP. It is encoded by an unique gene (At1g20050) possessing 4 exons and 3 introns.

Δ^7-sterol-C5(6)-desaturase. Δ^7-Sterol products of the Δ^8-Δ^7-sterol isomerase are transformed into Δ^5-sterols through two reactions : the first one involves a desaturation step at C5 leading to a Δ^5,7-sterol, then the Δ^7-double bond of Δ^5,7-sterols is reduced to give Δ^5-sterols (Fig. 2). Insertion of the 5,6-double bond has been shown to involve stereospecific removal of the 5 (4-proR in MVA) and 6 (5-proS in MVA) hydrogen atoms during ergosterol biosynthesis in *S. cerevisiae*.

The same stereospecificity is seen during the insertion of the 5,6-double bond during sterol biosynthesis in rat liver, higher plants and *Ochromonas malhamensis*. The close similarity of fatty acid desaturation in rat liver microsomes to the introduction of the sterol 5,6-double bond has been pointed out. Both systems require NADH and molecular oxygen, and involve a cyanide-sensitive factor; they both abstract cis-oriented hydrogens, occur in the microsomes and are inhibited by cyanide and thiol-blocking agents, but not by carbon monoxide.

Further evidence of this came witth the demonstration of the participation of cytochrome b5 in the conversion of cholest-7-en-3-ol into cholesta-5,7-dien-3-ol. The characterization of a Δ^7-sterol C5(6)-desaturase from *Zea mays* has been reported. This desaturase presents features similar to those reported for rat liver and yeast. It is strongly inhibited by cyanide, is sensitive to 1,10-phenanthroline and to salicylhydroxamic acid, but is insensitive to carbon monoxide, thereby suggesting the involvment of a metal ion, presumably iron, in an enzyme-bound form in the desaturing system. More recently the *ERG3* gene from *S. cerevisiae* has been cloned by complementation of an *erg3* mutant defective in C5-

sterol desaturase. The functional gene contained an open reading frame of 365 amino acids.

Gene disruption demonstrated that *ERG3* is not essential for cell viability. A nuclear and recessive mutant of Arabidopsis affected in sterol biosynthesis (*ste1*) has been isolated and identified. This mutant accumulated Δ^7-sterols (24R)-24-ethyl-5α-cholest-7-en-3-ol, and (24)-24-methyl-5-cholest-7-en-3β-ol at the expense of the normally occurring campesterol and sitosterol. It has been suggested that *ste1* is defective in the sterol-C5-desat-urase.

To definitely identify *ste1*, homologous complementation was necessary. To this end an Arabidopsis cDNA encoding a Δ^7-sterol-C5-desat-urase was isolated and characterized by functional complementation of *erg3*. This was achieved by transformation of *erg3* with an Arabidopsis cDNA library inserted in a yeast vector (pFL61). Transformants were screened for cycloheximide resistance and resistant clones were analyzed to determine their sterol profile.

The presence of ergosterol was detected in one clone in which a plasmid containing a cDNA of 1141bp was found. The 1141 cDNA allowed the deduction of an open reading frame of 843 bp encoding a 281 amino acid polypeptide. Three histidine-rich motifs (HX3H, HX2HH and HX2HH) and three transmembrane segment were found in the Arabidopsis polypeptide and are also present in Erg3p as well as in *Nicotiana tabacum* and *Homo sapiens* orthologues. Histidine-rich motifs are also characteristic of many membrane-bound fatty acid desaturases from higher plants. Overexpression of the Arabidopsis desaturase cDNA driven by a 35S promoter in transgenic *ste1* plants led to full complementation of the mutant. This result demonstrated that *STE1* was the impaired component in the desaturation system. The *ste1*-

derived open reading frame has been cloned by RT-PCR. Alignment of the wild-type with the *ste1* derived protein sequences revealed a single amino acid substitution : T114 in the wild-type is changed to I114 in Ste1p. The presence of this mutation in the mutant *STE1* genomic sequence demonstrated that the change of the T114 to I was necessary and sufficient to create the leaky allele *ste1*.

The role of 15 residues in the reaction catalyzed by Arabidopsis Δ^7-sterol-C5(6)-desaturase (5-DES) was investigated using site-directed mutagenesis and expression of the mutated enzymes in an *erg3 S. cerevisiae* strain defective in 5-DES. One group of mutants was affected in the eight invariant histidine residues from three histidine-rich motifs. Replacement of these residues by leucine completely eliminated the desaturase enzymatic activity both *in vivo* and *in vitro* in agreement with the hypothesis that the function of histidine-rich motifs would be to provide the ligands for a presumed catalytic Fe center, as previously proposed.

Another group of mutants was affected in residue 114 based on previous observations indicating that mutant T114I (*ste1*) was deficient in 5-DES. It was shown that the enzyme T114I had an 8-fold higher Km and 10-fold reduced catalytic efficiency. Conversely, the functionally conservative substituted mutant enzyme T114S displayed a 28-fold higher Vmax value and an 8-fold higher Km value than the wild-type enzyme. The data suggested that T114 strongly affected the reactivity of the desaturase and gave clues to get engineered 5-DES with much higher performances. Deuterated 7-cholestenol analogues were used as mechanistic probes for wild-type and mutated Δ^7-sterol-C5(6)-desaturase. Deuterium kinetic isotopic effects for the removal of 6α- and 5α-hydrogens were measured.

The Arabidopsis 5-DES considered above is encoded by one gene (At3g02580). Sequencing of Arabidopsis genome has revealed the existence of a second gene (At3g02590), the deduced protein sequence of it having 80% identity with Ste1p. Both genes possess three exons and two introns. Screening of about 50,000 M2 seeds of an EMS mutant population allowed the isolation of 43 dwarf mutants. Two of them (*dwf7-1* and *dwf7-2*) were biochemically complemented by brassinolide (BR) and early BR biosynthetic precursors.

Sequencing revealed a premature stop codon in exon1 (*dwf7-2*) and in exon 3 (*dwf7-1*) of the the *5-DES* gene confirming that *dwf7-1* and *dwf7-2* were null alleles of *STE1*. An extremely dwarf mutant (*bul1-1*) of Arabidopsis was recently isolated and shown to contain as much as 90% of Δ^7-sterols and less than 2% of Δ^5-sterols. It was therefore defective in the Δ^7-sterol-C5-desaturation step leading to brassinosteroid biosynthesis. Cellular characterization and rescue experiments with brassinosteroids demonstrated the involvment of the 5-DES in brassinosteroid-dependent plant growth processes.

Indirect immunofluorescence of a-tubulin in *bul1-1* revealed a total lack of the parallel microtubule organization that is typical of elongating cells in the wild-type. Rescue experiments with brassi-nosteroids and subsequent molecular analyses suggested that a brassinosteroid-responsive pathway exists, which allows microtubule nucleation/organization and cell elongation without activation of tubulin gene expression.

Δ^5,7-sterol Δ^7-reductase (Δ^7-SR) catalyzes the reduction of the Δ^7-double bond of the Δ^5,7-sterols into Δ^5-sterols. This step occurs only in vertebrates and higher plants. A microsomal preparation from seedlings of *Zea mays* catalyzed the NADPH dependent reduction of the Δ^7-bond of Δ^5,7-cholestadienol.

Novel 6-aza-B-homosteroids were shown to strongly inhibit (Δ^7-SR) and to behave as analogues of a high energy carbocationic intermediate occurring during the reaction. The gene encoding the Δ^7-SR was cloned by transformation of wild-type *S. cerevisiae* with a cDNA library of Arabidopsis constructed in the yeast vector pFL61 and the subsequent selection of cells producing Δ^5-sterols at the expense of ergosterol (a Δ^5,7-sterol) by nystatin selection.

These cells were selected due to their resistance to nystatin, a polyene fungicide that is highly toxic to ergosterol producing cells. This clever strategy allowed recovery of a plasmid bearing a 1290-bp cDNA open reading frame, which encoded a protein of 430 amino acids that had significant sequence identity (32%) with Arabidopsis sterol Δ^{14}-reductase and the previously reported LLXSGWWGXXRH signature of sterol reductases.

Cholesta-5,7-dien-3β-ol was shown to be reduced in vitro to cholesterol by an enzymic preparation from a yeast strain expressing the Arabidopsis (Δ^7-SR). Screening of a population of dwarf mutants led to a series of mutants (*dwf5*) presenting a defect in the gene encoding the Δ^7-SR. *dwf5* plants display the characteristic dwarf phenotype typical of BR mutants. This phenotype included small, round, dark-green leaves, and short stems, pedicels, and petioles. *dwf5* mutants were biochemically complemented with exogenous BRs. Metabolite tracing with ^{13}C-labeled precursors in *dwf5* and sterol analysis suggested a deficiency in a Δ^7-SR activity. In particular sterol analysis showed that $\Delta^{5,7}$-campestenol (a possible substrate of Δ^7-SR) accumulated. Surprisingly, the major compound analyzed was shown to be 5a-ergost-7-en 3β-ol. A Δ^5-reductase (*det2*) acting downstream of campesterol could reduce the Δ^5 double bond of $\Delta^{5,7}$-sterols which accumulated in *dwf5*.

Deficiency in Δ^7-SR activity was verified by DNA sequencing showing that all six independent alleles contained loss-of-function mutations in the Δ^7-SR gene. The *dwf5* plant could be restored to wild type by ectopic overexpression of the wild-type copy of the gene.

This gene (At1g50430) was shown to contain 13 exons and 12 introns. The Smith-Lemli-Opitz syndrome (SLOS) is an inborn disorder of sterol metabolism. All patients suffer from mental retardation. The SLOS gene has been shown to be a Δ^7-SR required for the *de novo* cholesterol biosynthesis. The protein encoded by the *Homo sapiens* Δ^7-SR has 35% identity with the Arabidopsis Δ^7-SR.

Δ^5-sterol Δ^{24}-reductase(isomerase) catalyzes the reduction of the Δ^{24} double bond of the side chain of sterols. This reaction uses a substrate possessing one double bond in animals and in higher plants. In animals this substrate is desmosterol, a $\Delta^{24}(25)$ sterol, whereas in higher plants, substrates are 24-methylene cholesterol and isofucosterol (Fig. 2) and $\Delta^{24}(24^1)$-sterols are probably isomerized in $\Delta^{24}(25)$-sterols prior to be reduced.

However, reduction of the Δ^{24} double bond proceeds on sterols possessing two conjugated double bonds such as ergosta-5,7,22,24-tetraen-3β-ol in *S. cerevisiae* and other organisms producing ergosterol. The *ERG4* gene provides the enzyme that reduces the double bond at C24(24^1), the final step in ergosterol biosynthesis. Disruption of *ERG4* led to a mutant (*erg4*) which was able to grow similarly to the wild-type showing that the C24 reductase was not essential for viability. The sterols accumulated in *erg4* mutants (e.g. ergosta-5,7,22,24-tetraen-3β-ol) are similar in structure to ergosterol and could fulfil functions normally attributed tc ergosterol.

The *diminuto* mutant (*dim*) was initially isolated as a slowly growing dwarf Arabidopsis mutant with greatly reduced fertility. The *DIM* gene was cloned, and genomic analysis and gene expression studies strongly suggested that *dim* did not produce any DIM protein. The *DIM* gene (AT3g19820), also referred to as *DWF1* or *CBB1*, is composed of two exons and one intron. Both the mutant phenotype and gene expression could be rescued by the addition of exogenous BRs.

Analysis of endogenous sterols demonstrated that *dim* accumulated 24-methylene cholesterol but was deficient in campesterol, an earlier precursor of BR and in BRs as well. Feeding experiments using deuterium-labeled 24-methyl-enecholesterol and 24-methyl desmosterol confirmed that *DIM/DWF1* is involved in both the isomerization and reduction of the $24(24^1)$ bond and encoded a sterol $C24(24^1)$ reductase isomerase (S24REDISO).

Transient expression of a green fluorescent protein-DIM/DWF1 fusion protein and biochemical experiments showed that DIM/DWF1 is an integral membrane protein that most probably is associated with the ER. Several *dwf1* alleles were also characterized. 7 of 10 *dwf1* mutations were shown to directly affect a flavin adenin dinucleotide binding domain conserved in various oxidoreductases. The proteic sequence of S24REDISO from Arabidopsis was shown to have 41% identity with an *Homo sapiens* ortholog (seladin-1) but no significant identity with ERG4. No ortholog of S24REDI has been reported in the *S. cerevisiae* genome. Thus the C24 reduction step seems to be performed by completely different enzymatic systems in higher plants and animals on one hand and in *S. cerevisiae* (and probably most fungi) on the other hand. The human *DIM/DWARF1* homolog seladin-1 has been shown to confer resistance to Alzheimer's disease-associated neurodegeneration and oxidative stress.

This exciting result would attribute a novel role of cholesterol in neurodegenerative diseases. However the involvment of seladin-1 in cholesterol biosynthesis has not been demonstrated so far. Finally mutations in the 3β-hydroxysterol Δ^{24}-reductase gene were shown to cause desmosterolosis, an autosomal recessive disorder of cholesterol biosynthesis.

Sterol-Δ^{22}-desaturase. Ergosterol from *S. cerevisiae*, most fungi and several algae, and stigmasterol : (24S)-24-ethyl-cholesta-5, E-22-dien-3β-ol from most higher plants possess a double bond at C-22. Δ^{22}-sterols are not present in vertebrates. This double bond results from a desaturation step. This step is the antepenultimate during ergosterol biosynthesis, it is probably the last step of plant sterol biosynthesis. The substrate of Δ^{22}-desaturase is ergosta-5,7,24(24^1)-trien-3β-ol in yeast, which is transformed into ergosta-5,7,22,24(24^1)-tetraen-3-ol by yeast microsomes.

This reaction has been shown to involve molecular oxygen, NADPH, and is inhibited by CO and metyrapone. Thus it involves a cytchrome P-450 species. The gene (*ERG5*) encoding this reaction has been recently cloned by complementation of an *erg5* mutant. The deduced amino acid sequence (538 amino acid residues) presented general features common to most cytochrome P-450 open reading frames and exhibits the highest homology with some animal cytochrome P-450 proteins of the family 2. Surprisingly, the highest homology was not with CYP51A1 (lanosterol-14-demethylase). Thus *ERG5* is the first member of the *CYP61* family.

Most biosynthetic studies show that sitosterol would be the substrate for a C-22 desaturase leading to stigmasterol. Very little is known about this step in higher plants, since the *in vitro* conversion of sitosterol to stigmasterol has never been shown enzymatically. Arabidopsis is remarkable among higher plants since it was shown to contain brassicasterol (ergosta-5, E-22-dien-3-ol) in addition to stigmasterol, indicating that both 24-methyl and 24-ethyl sterols could be substrates of the C-22 desaturase.

10

Functions in Secondary Metabolites

In the plant kingdom do ecology, evolution, and human affairs come together as intricately as in phenomena involving secondary metabolites. Secondary metabolites are those compounds produced by plants that are not directly essential for basic photosynthetic or respiratory metabolism; such compounds are known as primary metabolites. The importance of these compounds is apparent in some of the earliest annals of human history.

As a testament to the importance of secondary metabolites, P. somniferum was one the first plants to be cultivated for a reason other than its caloric or nutritive value. Sadly, its importance has only increased over time. Organic chemists of the nineteenth century were among the first to recognise and study the diversity of plant secondary metabolites and to analyse them both in terms of structure and function. The German organic chemists developed techniques to acetylate morphine and produce what they claimed at the time to be a nonaddictive version of the drug. Soon, however, the company realised that their new product, sold under the trade name Heroin, was vastly more addictive.

In another striking example of profitable secondary metabolites, similar acetylation techniques to salicylic acid, derived originally from the bark of willows, Salix, to create aspirin. The modern appreciation of these secondary metabolites from ecological and evolutionary perspectives had its spread outside of the plant-insect ecology community with Fraenkel's 1959 article in Science, "Raison d'E^ tre of Secondary Plant Substances." In this article he postulated that the overwhelming array of chemicals found in plants were more than simply by-products of primary metabolism that served as waste products for these organisms that lacked excretory systems. By the time of Fraenkel's article, the dust had largely settled on the question of whether nonnutritive factors such as attractants or repellents played a major role in insect host preference.

Fraenkel, Dethier, and their colleagues had provided overwhelming evidence that insects paid attention to these secondary substances in making their choices about what plants to eat. Ecologists and evolutionary biologists soon took up the cudgels, and the past 40 years have seen the development, testing, refinement, and, in some cases, rejection of models that describe the ecology and evolution of these "secondary" substances.

With the functional roles of these compounds broadly acknowledged, several challenges presented themselves to scientists interested in their ecological and evolutionary dynamics. Most obviously, the details of the processes underlying the interactions between each of these compounds and their target pest or pathogen became (and remain) one of the most vibrant topics in ecological and physiological research.

Two features of these chemicals and the interactions between the plants that produce them and their respective pests raised questions that continue today to motivate ecologists, evolutionists, agriculturalists, and legions of graduate students. First of all, the staggering diversity of "secondary" chemicals found in plants, even in closely related taxa, immediately raised questions regarding the origin and maintenance of this diversity. Second, the role of these chemicals in modulating plant-pest interactions and the prevalence of host specificity among phytophagous insects are poorly understood but essential issues for ecological and evolutionary studies as well as for applied questions in agriculture.

Evolutionary and ecological studies of plant secondary metabolites have been closely linked because of the early recognition that many of these compounds played distinct ecological roles while also showing phylogenetic affinities, such that details of the distribution of chemical structures among taxa could not be explained without considering evolutionary relationships. However, it soon became apparent that phylogenetic relationships among plants were not sufficient to predict functional roles of secondary compounds. For example, chemically similar secondary metabolites such as terpenes can serve a range of functions, including herbivore deterrence, fungal toxicity, and pollinator attraction; they can also function as solvents for higher molecular mass compounds that would otherwise solidify and clog transport systems in plants.

Ecological and evolutionary models have focused on the origin and maintenance of plant chemical diversity and on the regulation of chemical concentrations within plants. There has been less effort to date on formal models of the evolution of secondary metabolite function and of how function may change over time, though some researchers have examined these questions on the scale of particular plant clades. Recent advances at the genetic, physiological, and ecological scales allow us to use terpenes as models for the study of the evolution of chemical function and of the ecological impacts of evolutionary change.

Evolutionary Models

With the recognition that secondary metabolites were not just plant waste products came attempts to understand their distribution and abundance. Two challenges immediately faced efforts to develop models of allocation to secondary metabolites. The first was to understand the origin and maintenance of the astounding diversity of chemicals produced in the plant kingdom, and the second was to discover the patterns underlying the variation in quantities that are found both within and among plants of the same and different species. The diversity of chemicals found within plants has been explained as resulting from the interaction of plants and their major pests, pathogens, and predators.

Ehrlich and Raven proposed that the evolution of secondary plant metabolites and the stepwise evolutionary responses to these compounds by phytophagous insects have led to the radiation and diversity of both of these groups. While the stepwise process of escape and radiate has come under close scrutiny, the complexity of this battle between plants and insects is well documented in the diversity of secondary phytochemicals that act as deterrents against herbivores. The model of Jones and Firn brings advances in our understanding of chemical activity to bear on the question of chemical diversity and indicates that plants will have a much higher diversity of compounds than are used for defense at any one time.

Together, the models of Jones and Firn and Ehrlich and Raven explain the amasing chemical diversity observed. These ecological and physiochemical models must, however, be evaluated within a phylogenetic framework when attempts are made to apply them to empirical data. The more challenging issue has been, and continues to be, developing a model for predicting the quantities of secondary metabolites found in plants. The work of Herbert Simon to extend optimality modeling and linear programming techniques from economics to other disciplines soon found a receptive home in ecology and evolutionary biology.

Thus, in evolutionary terms, there must be a cost to defense, and this cost can be calculated by estimating the foregone reproduction resulting from allocation to defense. The benefit of allocation to defense is then the increase in reproduction that occurs as a result of a plant's not losing biomass to pests or pathogens. Coincident with this explicit optimality approach, Feeny developed a model of allocation to secondary metabolites that was based on a notion of the risk faced by plants. This model, the apparency model, partitions plants into those that are likely to be specialised on by herbivores (plants that are apparent) in contrast with those that are not.

The "unapparent" plants, ephemeral in their distribution, are more likely to evade detection by specialists and can afford defenses that work at low concentrations and are mobilised for reuse. The concept of apparency brings explicit consideration of risk to the calculation of optimal levels of defense in plants. In the early 1980s attention began to focus on the role of resource availability as a parameter affecting growth rate and the relative costs and benefits of producing secondary metabolites for defense. This optimality approach is seen in the evolutionary model of Coley et al. and Bloom et al. A decade ago Sangerl and Bazaz brought together risk and resource availability into a comprehensive optimality model that has proven extremely effective at predicting allocation to defense.

Among the models not based on optimality principles, perhaps the most significant model was put forth in the early 1980s by John Bryant et al. They proposed that allocation to various defensive compounds was governed by basic stoichiometric requirements and that when resources were available in excess of a plant's ability to use them for growth or reproduction, the resources would be allocated to secondary metabolites. This model, termed the carbon-nutrient balance hypothesis, has excited much heat and a little light, but its intellectual status is more akin to the sort of null model that all good optimality models must posit simultaneously than to a general challenge to optimality thinking.

The challenge now facing ecologists and evolutionists is less to discern which particular optimality model is most applicable to particular plants and their secondary metabolites than it is to bring together advances in the genetic and physiological bases of secondary metabolite function with ecological and phylogenetic patterns of compound distribution. For terpenes enough is known, to begin to draw these connections and develop a theoretical perspective that uses ecology and evolution to understand diversity and abundance from the genetic to the phylogenetic scale and that takes into account the functional roles of these compounds.

Biochemistry of Terpenoids

The functional diversity of chemicals within plants is best demonstrated by terpenoids. More than 30,000 terpenoids have been identified, each one composed of 5-carbon (prenyl diphosphate) building blocks added together to make 10-carbon, 15-carbon, and even 2000- 500,000-carbon chains, as is the case for rubber. Terpenes are functionally diverse, and many are integral to primary metabolism, e.g., hormones such as gibberellins and abscisic acid, electron carriers such as plastoquinone and ubiquinone, terpene-derived compounds that form structural parts of membranes such as phytosterols, and photosynthetic pigments such as carotenoids.

There are two biochemical pathways responsible for the synthesis of terpenoids. In both prenyltransferases act in the condensation of an isoprene unit, IDP (isopentenyl diphosphate), with an IDP isomer, DMADP (dimethallyl diphosphate), in a head-to-tail addition. The result, prenyl diphosphates of varying lengths, is then modified through dimerisation and cyclisation by TPSs (terpene synthases) to produce a diverse array of terpenoids with an almost equally diverse array of functions. The head-to-tail addition of isoprenoid units was initially identified as the biosynthetic mechanism for the synthesis of terpenoids. The first pathway described, one that is common to bacteria, fungi, mammals, and plants, is the mevalonic acid pathway.

In plants there are two subcellular locations for the synthesis of terpenes, the cytosol and the plastids. It was not until 1997 that Lichtenthaler et al. discovered that the plastidbound pathway was distinct biochemically. They showed that this "new" pathway, the DOXP/MEP (deoxyxylulose phosphate/ methylerythritol) pathway, is identical to that discovered in bacteria a few years earlier and probably is a legacy of prokaryotic endosymbiotic ancestors. Thus, in higher plants there are two pathways for generating terpenes. The classic mevalonic acid pathway produces the precursors of sesquiterpenes (C15) and multiples thereof, for example, the triterpenes (C30), via the cytosol; and the plastidic DOXP/MEP produces units of C10, such as the precursors of monoterpenes (C10) and diterpenes (C20), etc.

Recent evidence indicates that there may well be sharing of intermediates across these pathways, a sort of biosynthetic cross talk. A common ancestor has been implicated for both prokaryotic and eukaryotic prenyltransferases. These enzymes share two highly conserved regions that contain the aspartate-rich consensus sequence DDXX(XX)D. This sequence has been postulated as the binding site for the metal ion-complexed diphosphate group. The electrophilic reaction mechanism proceeds through the generation of a carbocation intermediate. This mechanism is also used by the TPSs, which share the similar DDXXD motif. These electrophilic reactions in TPSs and the manner in which the resulting carbocation is quenched give rise to the great diversity of terpenoids functioning in so many different metabolic processes.

Evidence for the common heritage of the plant terpene synthases comes from the similar reaction mechanism, as well as from sequence similarity, including exon and intron size and conserved sequence motifs. Phylogenetic analysis of 33 TPSs from angiosperms and gymnosperms established that the origin of TPSs occurred before the splitting off of the angiosperms. Synthases from angiosperms and gymnosperms functioning in primary metabolism are more closely related than angiosperm synthases are to each other. However, the specific functions of monoterpene, sesquiterpene, and diterpene synthases involved with secondary metabolism evolved after the divergance of the gymnosperms and the angiosperms. For example, the limonene synthases in gymnosperms (Abies) and angiosperms (Mentha) are not closely related but are, instead, an example of convergent evolution.

S-linalool synthase, an ensyme that produces S-linalool, an attractive component of some floral blends, is structurally the most distinct synthase gene identified so far. This distinctiveness may not be surprising as the action of this ensyme is different from other synthases, catalysing a simple ionisation, with water trapping the carbocation rather than the inter-or intracarbon carbon bond formation. However, a common origin between plant-linalool synthase and bacterial-linalool synthase is unlikely, since it has been hypothesised that in plants this synthase is actually a composite gene that evolved through a recombination event between a copalyl pyrophosphate synthase-like gene and the limonene synthase-like gene.

In general, the synthases involved in the production of terpenoid secondary metabolites are far more readily adaptable than synthases involved in the production of terpenoid primary metabolites. While gene duplication is believed to be the primary mode for evolving a new trait from an old gene, the evolution of function in secondary metabolites may not require doubling; if a gene is not critical for organismal function, there may be continuous opportunity for new functions to evolve by mutation, leading to higher diversity.

When such a change occurs in an isolated population it may be tied to cladogenesis. In addition, functional diversity can arise with very little change in synthase structure. Closely related genes can encode ensymes synthesising dissimilar monoterpenes and sesquiterpenes. Bohlmann et al. showed that TPSs with greater than 70%-90% identity can catalyse distinct chemical reactions, whereas other TPSs with no more than 30% similarity may catalyse the same reaction. By swapping regions on two synthase genes, 5-epi-aristolochene and vetispiradiene, Back and Chappell were able to demonstrate that just one exon is responsible for substrate specificity while another exon is responsible for the different reaction products.

Another path to diversifi-cation involves ensymes that synthesise multiple products depending on protein environment. In some cases only a single product is formed, while in other cases numerous products are generated by the same synthase. For example, in grand fir, Abies grandis, the substrate farnesyl diphosphate (FPP) leads to an extraordinary diversity of end products when acted on by constitutive synthases such as γ-humulene synthase (three major and 52 minor sesquiterpenes) or δselinene synthase (three major and 34 minor sesquiterpenes).

In contrast, inducible synthase ensymes such as d-cadinene synthase and (E)-a-bisabolene synthase each produce a single end-product. These two single-product ensymes are induced in response to herbivore damage, suggesting perhaps a general pattern of constitutive terpene synthases leading to multiple end-products, while those that are induced in response to damage have fewer end-products. Clearly, further work must be done to test the generality of this pattern. In terms of the factors regulating the quantity of terpenes produced, the primary regulation appears to reside in the first step of the pathway, the conversion of prenyl diphosphate into the structurally diverse higher terpenes.

From studies on isoprene synthase in velvet bean, linalool synthase in Clarkia breweri, and limonene synthases in mint, levels of terpene synthase activity have been correlated with the production of terpenoids, although levels of precursors could ultimately be a rate-limiting step. For example, the level of monoterpenes in the oil glands of mint species reflects the enzyme levels. Similarly, linalool synthase activity closely matches the production of linalool and its derivatives. However, when petunia was transformed with Slinalool synthase cDNA linalool was successfully expressed but not emitted. Instead it was converted to the nonvolatile -linalylb-D-glucopyranoside.

Induction by wounding or by fungal elicitors may also lead to an upregulation of TPSs and prenyltransferases. However, with the notable exception of Mentha, little is known about the ensymes along the pathways. Without information on concentration of intermediates and flux rates through the pathway, it is difficult to accurately assess the most important regulatory steps. Genetic variation (both within and among species) occurs in the context of both epistatic interactions and simple Mendelian relationships.

For example, epistatic interactions have been implicated for the two best-studied systems among the angiosperms, Thymus and Mentha. In Thymus vulgaris there are six chemotypes in which epistatic interactions form a dominance hierarchy. The chemotypes, in rank order of dominance, are geraniol 1 alpha terpineol 1 thuyanol-4 1 linalool 1 carvacrol 1 thymol. A series of six loci each with a dominant and a recessive allele controls the branching along the pathway and, thereby, the chemotype produced. Unless the plant is homosygous recessive for geraniol, geraniol will be produced; if it is homosygous recessive for geraniol, then alpha terpineol will be produced; if it homosygous recessive for both geraniol and alpha terpineol, then thuyanol-4 will be produced, and so on, along the hierarchy of the six morphs.

In Mentha different species have been extensively studied to determine the genetic control of the monoterpenes in this genus, with attention paid both to the end-products and to the ensymes that produce the intermediates. Most of the genes in the pathway appear to be diallelic with simple dominant and recessive alleles. For example, limonene synthase is controlled by two diallelic genes that interact epistatically. The interactions between these two genes control the production of the three possible end-products: (-)-*trans* carveol, (-)-*trans*-isopiperitenol, or limonene itself. Complicating the picture further, a third allele epistatically affects these two genes and can lead to the synthesis of linalool and linalyl acetate rather then any of these three cyclic monoterpenes.

There are a number of examples of simple genetic control by one major dominant gene or by closely linked loci for angiosperms as well as gymnosperms. In Perilla frutescens, there are two dominant genes essential for the formation of cyclic monoterpenoids, and in Achillea simple dominant genetic control has been identified for the sesquiterpene longipinenone. In Pinus elliotii, both bpinene and myrcene are each controlled by single but independent genes. Similarly, δ-3-carene, a monoterpene present in Pinus pinaster, is controlled by one major gene or closely linked loci.

Experimental studies using hybrids have produced similar results. Raguso and Pichersky found that S-linalool in C. breweri was controlled by simple Mendelian dominance. Shepherd et al. used QTL analysis on half sibs and also found the monoterpenes of a eucalypt hybrid to be controlled by a single gene region. In short, it is difficult to generalise the relative importance of simple Mendelian compared with complex epistatic relationships in determining the composition and quantities of terpenes produced. The apparent ubiquity of both mechanisms, however, indicates multiple possibilities for genetic control, and, thus, evolutionary change, in most taxa.

Function of Terpenoid Production

Terpenoid secondary metabolites occur across a wide range of plant tissue types; those serving as defensive chemicals are often secured in secretory structures. Such specialised structures minimise the risk of autotoxicity but maintain terpene concentrations at sites whose defense is crucial. In conifers, terpenoids are stored in resin ducts or lactifers under pressure, which can have a devastating impact on small herbivores. In angiosperms, glandular trichomes that contain terpenes often serve as a first line of defense against would-be herbivores. These externally based structures enhance the effectiveness of chemical defenses because they can arrest feeding preemptively. In addition, pellucid glands and secretory ducts within leaves are also common.

In conifers, oleoresin, a complex mixture of terpenoids that includes monoterpenes, sesquiterpenes, and diterpenes, is an important defense strategy against bark beetles and their associated fungal pathogens. There is great diversity in the manner of storage and release for this defensive blend. On one end of the continuum are compounds stored in branched interconnected canals, e.g., the resin canal system of Pinus. Attack by an herbivore or pathogen causes very little biochemical induction of terpene synthesis in conifers with complex canal systems; the main plant response involves the transport of terpenes through the canals to the site of attack, often across several meters. At the other end of the spectrum are the simple resin cells of Thuja (cedar), which can be induced by attack but are normally either absent or present in very low quantities. Abies (true firs) have multicellular resin blisters that are less complex in Larix, Picea, and Pseudotsuga (larch, spruce, and Douglas fir), where the blisters are connected but also inducible.

As Gershenzon and Croteau point out, the defense theory proposed by McKey and Rhoades of "risk and value" has largely proven useful in locating the tissues with the highest levels of terpenes. Concentrations of terpenoids are generally higher in reproductive structures and typically highest in the foliage during and immediately following flowering. Young leaves are more important to a plant than older leaves, and the highest levels of terpenoids are found in the young organs. Variation in concentration can also vary within a tissue. The pattern of constitutive monoterpenes along the needles of ponderosa pine slows caterpillar feeding. Terpenes increase as the caterpillar feeds, motivating the caterpillar to switch needles and slowing the rate of damage, thereby giving the plant time to induce TPSs. For terpenoids that are released to attract insects, specialised storage structures are far less common.

These compounds are generally released as they are produced and so are maintained within plant tissues far below toxic levels. However, there are exceptions to this pattern. Osmophores are specialised scentproducing organs of variable morphology and ontogeny, many of which store the volatiles they produce. In orchids such as Stanhopea, Clowesia, and Polycycnis, euglossine bees are attracted to the terpene-based scent that is emitted from these structures. Specialised structures are not necessary for the establishment of spatial patterns in floral emissions. Pollen and anthers can have altogether different mixtures of volatile scents compared to the rest of the flower, and many pollen-feeding insects are adapted to discerning these scent mixtures.

Scent guides that bolster visual nectar guides have also been identified, indicating that interactions among sensory modalities may be crucial in determining the function of particular terpenes. Dalechampia is an excellent example of the crucial role that spatial patterns of scent emission can play in plant-insect interactions. From a phylogenetic study, Armbruster et al. determined that a triterpene resin underwent an evolutionary switch in function from a deterrent and toxin to an attractant and reward, and then back again to its deterrent role.

Ancestrally, resin functioned as a defense, evenly distributed in the floral tissue. However, in a group of more derived species within the same genus, pollinated by resin-collecting bees, resin ducts are unified and positioned for optimal pollinator attraction leading to greater pollination efficiency. In the most derived species, resin ducts occur in the leaves and protect the foliage from herbivores. The chemical composition of the resins is similar in all three taxa; it is the placement of the resin duct within the plant that defines its functions.

Evolution of Function in Terpenoids

For all compounds involved with interorganismal signaling, function can evolve through changes in the recipient organisms rather than through changes in the sending organism. These toxic effects have been demonstrated mostly for generalist herbivores, but many of these studies have been conducted on agricultural pests rather than actual herbivores the plant is likely to encounter in a natural setting. Many specialist insects, which the plant certainly does encounter, have seemingly adapted to these defensive compounds. No longer do they find these compounds repellent; instead the compounds may be used as cues for host recognition, even when they retain their growth inhibiting effects. While feeding on the toxin may reduce growth, fitness can actually increase through reduced predation due to sequestered toxins.

Few taxonomic groups have been put to the test, because a sound test requires both a robust phylogeny and ample ecological information. Most of the research conducted on plant-insect interactions has not explicitly addressed even reciprocal adaptation, much less the "escape and radiate" hypothesis of Erhlich and Raven.

In the few cases where it has been properly tested, the hypothesis has been both rejected, e.g., Mexican flea beetles, Blepharida, on its host Bursera, and supported, e.g., Phyllobrotica beetles and plants in the Lamiales and Tetraopes (longhorned beetle) on Asclepias (milkweed). This last example appears to follow a phylogenetic progression in the potency of the cardenolides, tracked by the counteradaptations of the beetle lineage. These linked changes are especially intriguing as the beetles are just one of the many groups of herbivores that feed on Asclepias.

The most charismatic herbivore on Asclepias is certainly the monarch butterfly Danaus plexippus, which is the textbook example of a specialist herbivore that feeds on a toxin (car-denolides) and gains advantage via sequestration. Interestingly, for first instar larvae, a high concentration of cardenolides actually reduced the rate of growth and survival. Another specialist on Asclepias, the weevil Rhyssomatus lineaticollis, is cryptically colored and does not sequester the toxin, although there are high concentrations of cardenolides in the pith where the larvae feed. The stems on which adults lay their eggs have relatively lower concentrations.

These two alternative tactics highlight the flexibility that exists when natural selection can act on two operators; the strategy of feeding on Asclepias leads to very different ecological tactics that, in turn, are correlated with extreme differences in morphology, physiology, and behavior. A number of herbivores recycle terpenoids, using them for protection without sequestering them; they secrete the terpenes, and this secretion is used to ward off predators. Some species of the plant genus Bursera produce a resin after insect damage; specialist herbivores in the genus Blepharida utilise this resin by adorning their backs with a regurgitant and an anal secretion that match the hosts' defense chemistry.

The weevil Oxyops vitiosa produces a mixed terpenoid secretion garnered from its host plant Melaleuca quinquenervia. By covering its integument with this secretion, the weevil repels generalist predators. Another beetle, Eurypedus nigrosignata, constructs fecal shields using terpenes found in their larval host plant Cordia curassavica. Similarly, larvae of the leaffeeding beetle Lema trilinea carry fecal shields that contain steroidal glycoalkaloids and saponins, which are modified from their host Solanum dulcamara.

For volatile compounds such as the monoterpenes and sesquiterpenes, it is likely that the concentration of the compound is related to its function. As noted above, terpenoids whose function is involved with reducing herbivory tend to be stored in secretory structures or organs, which usually contain terpenes at high concentrations, on the order of milligrams per gram. Terpenoids involved with pollinator attraction, however, tend to be located in parts of the plant where the resistance to flux is very low, and substantial flux rates can occur even at low concentrations, on the order of micrograms per gram.

Changes in function through changes in receiver underlie some of the most important events in the evolution of land plants. During the Carboniferous period, ca. 300 Ma, flowerlike structures and insects capable of pollination began to emerge. By the Cretaceous, 150 million years later, pollen and ovules supplemented the diets of some predatory insects. The chemical deterrents against pests and pathogens emitted by plants were co-opted by insects seeking either food or mates. The chemistry of floral scents is similar to nonfloral deterrents, and theory indicates that these early herbivorous insects inadvertently acted as pollinators; thus a novel selection pressure arose for chemical volatiles to be attractive.

In a study of the scent emitted by insect-pollinated cycads, representing an independent evolution of insect attractants, Pellmyr et al. found classes of attractant compounds that are chemically similar to those produced by taxa with deterrent volatiles. Phylogenetic analyses of floral scent have demonstrated that, while there is some conservation in the classes of compounds produced, there is also a high degree of homoplasy at the level of the specific compound. The great diversity of compounds is likely maintained to promote pollinator fidelity. When rewards are not forthcoming, the importance of these signals is especially apparent and pollinator attraction occurs by deceit.

For example, in Ophrys spp. (Orchidaceae) chemical mimics of specific insect pheromones are emitted by the orchids in order to attract male pollinators. Some Aroids attract pollinators seeking oviposition sites by emitting an odor of carrion or excrement. Specificity of floral scent can be beneficial by encouraging fidelity, as in the extreme case of the figs and their fig wasp pollinators. It is likely that each of the 750 species of fig emits a blend specific enough to attract its one species of fig wasp. In geonomoid palms, there is also evidence for maintenance of reproductive isolation through pollinator attraction, and such relationships have been shown for a variety of taxa, including but not limited to Ranunculaceae, Cimicifuga and Orchidaceae, Satyrium.

In what appears to be the most recently derived function of volatile terpenes, the emission of typical "floral" compounds is induced from the foliage of cotton, tobacco, and corn by the feeding of larval lepidoterans. Parasitoid wasps, in turn, utilise this signal to locate their larval lepidopteran hosts. Upon positive identification, the wasp will oviposit into the feeding caterpillar. The specificity of this signal has been demonstrated; Heliothis virescens and Helicoverpa sea feeding on corn each elicit the production of a unique ratio/blend of terpenoids, as well as lipoxygenase-derived volatile compounds. The difference in the two signals is distinguished by the specialist parasitoid Cardiochiles nigriceps and allows it to pinpoint its host, H. virescens.

Similarly, the parasitoid Cotesia glomerata uses the signals produced by Brassica oleraceae var. gemmifera (brussel sprouts), where the emission of green leaf volatiles as well as sesquiterpenes and cyclic monoterpenoids is systemically induced after feeding by Pieris brassicae caterpillars. One of the most exciting recent discoveries involving volatile terpenes and signaling to predators and parasitoids involves the apparently ubiquitous induction of homoterpenes in response to herbivore damage. Homoterpenes are C-11 and C-16 compounds produced by the removal of four carbons from sesquiterpenes and diterpenes.

In certain plants, predators and parasitoids have learned to use the induced release of these compounds as cues to locate prey/hosts. It appears that this induction may be primitive across higher plants but that the insect response may have been independently derived multiple times. This "cry for help" has been demonstrated for a number of different plants soliciting predators. Linalool is induced in Phaseolus lunatus, lima bean, following infestation by the her-bivorous mites Tetranychus urticae.

The herbivore is repelled by this compound but not so the predatory mite, Phytoseiulus persimilis. In another example, sesquiterpenes are induced by potato plants that have been damaged by the Colorado potato beetle, Leptinotarsa decemlineata, andWeissbecker et al. found the predaceous stink bug, Perillus bioculatus, which is attracted to damaged plants, to be responsive to these compounds. Predator cuing in on insect damage has also been demonstrated for conifers.

Oxygenated monoterpenes attract the bark beetle parasitoid Roptrocerus xylophagorum after insect damage by two herbivores, the southern pine beetles and the southern pine engraver, in Pinus taeda, loblolly pine. Similarly, in spruce the bark beetle parasitoid Rhopalicus tutela is attracted to bark beetle larvae "concealed" under the bark of spruce trees by the emitted volatiles. The induction of volatile compounds as a result of insect feeding sets up the possibility that neighboring plants could, in fact, use these signals, much as parasitoids do, to detect insect presence on neighboring plants. This intriguing idea described as "talking trees" has been met with much controversy since it was first introduced 20 years ago. Proper controls have been lacking in many studies that find plants responding to neighboring induction.

Recently, Karban et al. demonstrated in manipulated field conditions that when sagebrush, Artemesia tridentate, emits an epimer of methyl jasmonate after mechanical damage, neighboring Nicotiana attenuata individuals respond by inducing defense compounds. The methyl jasmonate produced by the damaged plants has been shown to increase resistance to herbivores. The neighboring wild tobacco plants, which grow in close proximity naturally, increased their synthesis of polyphenol oxidase, an enzyme putatively involved with defense, and these plants experienced less herbivory than control plants. This response disappeared when the air between plants was blocked but not when the soil was blocked.

Recently, in a lab-based study lima bean has demonstrated plant-to-plant communication through a volatile signal. In the presence of an herbivore-damaged plant, the genes of undamaged lima bean elevate the expression of several terpenoids. Arguments from physical and chemical transport principles, however, suggest that plant-to-plant communication by volatile signals is probably exceedingly uncommon metabolites.

11

Modern Biotechnology

Biotechnology will be a critical element in the development of crops, foods, and ingredients with traits with improved nutritional properties. Developing plants with these improved traits involves overcoming a variety of technical challenges inherent in metabolic engineering programmes. Both traditional plant breeding and biotechnology-based techniques are needed to produce plants with the desired quality traits. Continuing improvements in molecular and genomic technologies are contributing to the acceleration of product development. Some of these crops have already been approved and commercialized, whereas others are still in development.

Plant Metabolism

Plants are remarkable in their ability to synthesise a variety of organic compounds, such as vitamins, sugars, starches, fatty acids, and amino acids. As many as 80000 to 100000 different substances are synthesised in plants, including macronutrients (for example, proteins, carbohydrates, lipids [oils], and fiber), micronutrients (for example, vitamins and minerals), antinutrients (for example, compounds such as phytate that reduce bioavailability), allergens (for example, albumin), endogenous toxicants (for example, glycoalkaloids and cyanogenic glycosides), and other plant-specific compounds (some of which may have beneficial effects) that are significant to human and animal health. This plasticity is elegantly demonstrated in the way that plants respond to environmental stimuli such as pathogen attack.

Functional complexity begins with the exogenous signals perceived from the pathogen, continues with the mechanisms of signal perception and signal transduction, and results in extensive "reprogramming" of cellular metabolism, involving extensive changes in gene activity. Thus, pathogen defence entails a major shift in metabolic activity, rather than altered expression of a few unique, defence-related genes. The observed complexity serves as a paradigm of the flexibility and plasticity of plant metabolism. Many of these same metabolites have either positive or negative impacts on the nutritional characteristics of plants. For example, the shikimate pathway includes a number of phytochemicals that can have either good or bad effects. These compounds include phenylpropanoids, coumarins, stilbenes (some such as resveratrol are beneficial, while others such as kawain have negative effects), flavonoids, and tannins.

Nutritional Quality

The next generation of plants will focus on value-added output traits where valuable genes and metabolites will be identified and isolated, with some of the metabolites being produced in mass quantities for niche markets. The nutritionally improved crops in the current development pipeline will be well understood and well characterised from a compositional perspective as they undergo safety and nutritional assessment following existing regulations that are more than adequate to address any potential concerns. However, some of the more potentially beneficial modifications will require a more thorough understanding of plant metabolism and methods to achieve effective changes in the desired metabolic end points.

Although progress in dissecting metabolic pathways and our ability to modify gene expression in GM plants has been most impressive during the past 2 decades, attempts to use these tools to engineer plant metabolism have met with more limited success. Metabolic engineering typically involves the redirection of cellular activities by the modification of the ensymatic, transport, and regulatory functions of the cell using recombinant DNA (rDNA) and other techniques. Since the success of this approach hinges on the ability to change host metabolism, its continued development will depend critically on a far more sophisticated knowledge of plant metabolism, especially the nuances of interconnected cellular networks, than currently exists.

Although the ensymological sequences and intermediates of many metabolic pathways in a small number of well-studied organisms are known with some confidence, little is known in quantitative terms about the controls and integration of these pathways. The necessary knowledge also includes conceptual and technical approaches necessary to understand the integration and control of genetic, catalytic, and transport processes. Though there are notable exceptions, most successful attempts at metabolic engineering thus far have focused on modifying (positively or negatively) the expression of single genes (or a series of individual ensymatic steps) affecting pathways.

Generally, more success has been achieved when conversion or modification of an existing compound to another has been targeted than when an attempt has been made to significantly change flux through a pathway (for example, increasing the oleic acid concentration in canola oil, as will be discussed later). Attempts to modify storage proteins or secondary metabolic pathways have also been more successful than have alterations of primary and intermediary metabolism. Research to improve the nutritional quality of plants has historically been limited by a lack of basic knowledge of plant metabolism and the stimulating challenge of resolving complex interactions of thousands of metabolic pathways.

With the tools now being harnessed through the fields of genomics and bioinformatics, there is the potential to identify genes of value across species, phyla, and kingdoms. Through advances in proteomics, it is becoming possible to quantify simultaneously the levels of many individual proteins and to follow post-translational alterations that occur in pathways. Metabolomics allows the study of both primary and secondary metabolic pathways in an integrated fashion. With these evolving tools, a better understanding of global effects of metabolic engineering on metabolites, ensyme activities, and fluxes is beginning to be developed. The increase in our basic knowledge of plant metabolism during the coming decades will provide the tools necessary to modify more effectively the nutritional content of crops to have a positive effect on many aspects of human and animal health.

In addition to metabolic considerations, attention needs to be given to the site of synthesis and site of activity of an ensyme. Signal sequences or transit peptides coding sequences attached to introduced genes are not always sufficient to ensure appropriate targeting.

For example, charge and size of a protein may affect the efficiency of transportation into plastids. Another complexity found in biological systems is redundancy of pathways and the ability of plants to compensate as they often contain more than one enzyme capable of catalysing a similar reaction. A potential approach to counter some of these problems in metabolic engineering of pathways involves the manipulation of transcription factors that control networks of metabolism.

For example, expression of maize transcription factors C1 and R, which regulate production of flavonoids in maize aleurone layers, together under the control of a strong promoter resulted in a high accumulation rate of anthocyanins in Arabidopsis, presumably by activating the entire pathway. Such expression experiments hold promise as an effective tool for the determination of transcriptional regulatory networks for important biochemical pathways. In summary, metabolic engineers must not only understand the fundamental physiology of the process to be impacted, but also the level, timing, subcellular location, and tissue or organ specificity that will be required from a transgene to ensure successful manipulation of that physiology.

Gene expression can be modulated by numerous transcriptional and post transcriptional processes. Correctly choreographing these many variables is the element that makes metabolic engineering in plants so challenging. In conjunction with such increases in the understanding of plant metabolism, the challenge then remains to understand how components in the diet interact with human or animal metabolism to benefit their health and well-being. This challenge is at least as complex as the task of increasing or decreasing the amount of a specific protein, fatty acid, or other component of the plant itself. It is of little use producing a plant with a supposed nutritional benefit unless that benefit actually improves the health of humans or animals.

Metabolic engineering is generally defined as the redirection of one or more enzymatic reactions to improve the production of existing compounds, produce new compounds, or mediate the degradation of compounds. Substrate-product relationships in plant pathways were initially elucidated through the application of radiolabel tracer studies during the 1960s and 1970s. In the 1980s, with the advent of rDNA technology, tools such as cloning, promoter analysis, protein targeting, plant transformation, and biochemical genetics were developed.

The GM crops with improved agronomic traits presently being grown on more than 60 million ha around the world are a product of the application of these technologies to crop plants. These products provide benefits to the farmer and community in reducing insecticide and herbicide usage and increasing the ability of farmers to conserve soil and other resources. They generally involve the relatively simple task of adding a single gene or small number of genes to plants. These genes in the main function outside of the plant's primary metabolic processes and thus have little or no effect on the composition of the plants.

The more complex task lies in engineering metabolic pathways and plant metabolites. Significant progress has been made in recent years in the molecular dissection of plant metabolic pathways and in the use of cloned genes to engineer plant metabolism in ways that are more complex. Metabolic pathway modifications are complex, and the state of understanding of plant metabolism is sometimes insufficient to bridge the gap between the ability to clone, study, and modify individual genes and proteins and the understanding of how they are integrated into and affect the complex metabolic networks in plants.

Regulatory oversight of such products has been designed to detect such unexpected outcomes and to ensure that products from GM plants are safe before they are commercialized. Genomics-based strategies for gene discovery, coupled with high-throughput transformation processes and miniaturized automated analytical and functionality assays, have accelerated the identification of product candidates. Identifying rate-limiting steps in synthesis could provide targets for genetically engineering biochemical pathways to produce augmented amounts of compounds and new compounds.

Targeted expression will be used to channel metabolic flow into new pathways, while gene-silencing tools can reduce or eliminate undesirable compounds or traits, or switch off genes to increase desirable products. In addition, molecular marker-based breeding strategies have already been used to accelerate the process of introgressing trait genes into high-yielding germplasm for commercialisation.

Experimental Modification Pathways

Analysis of fluxes in metabolic pathways in response to an environmental or genetic manipulation can help identify rate-limiting steps. Traditional biochemical hallmarks of potential regulatory, or rate-controlling, enzymes are that they catalyze reactions and are regulated by appropriate effector molecules. The modification of enzymes of the carbon cycle to study their role in regulating pathway flux has provided some of the more interesting results from metabolic engineering studies in plants. For example, when the highly regulated

Calvin cycle enzymes, fructose-1, 6-bisphosphatase and phosphoribulokinase, were reduced 3- and 10-fold in activity, respectively, minor effects on the photosynthetic rate were observed.

In contrast, a minor degree of inhibition of plastid aldolase, which catalyses a reversible reaction and is not subject to allosteric regulation, led to significant decreases in photosynthetic rate and carbon partitioning (Haake and others 1998). Thus aldolase, an ensyme seemingly irrelevant in regulating pathway flux, was shown to have a major influence over the pathway. Understanding of the individual kinetic properties of such key ensymes may not always be sufficient to understand their wider role in central metabolism.

Functional Foods

In recent years, a new category called "functional foods" has appeared in the marketplace, and sales are growing quickly. For many, functional foods include not only those with added components that enhance their health claims but also include unsupplemented foods for which new health claims are recognized through the addition of a new product label. Functional foods are intended to appeal to consumers by offering potential health benefits that go beyond satisfying basic nutritional needs. These foods exploit the growing scientific evidence supporting the role of a diet containing certain types of foods or phytochemicals in the prevention and treatment of disease.

Epidemiological research has shown a positive association between dietary intake of food components found in fruits, vegetables, grains, fish oil, and legumes and their effect on chronic disease. In 1992, a review of 200 epidemiological studies showed that cancer risk in people consuming diets high in fruits and vegetables was only half that in those consuming low amounts of these foods. Functional food components have been associated with the prevention and/or treatment of at least 4 of the leading causes of death in the USA: cancer, diabetes, cardiovascular disease, and hypertension.

The U.S. National Cancer Institute estimates that 1 in 3 cancer deaths are diet related, and that 8 of 10 cancers have a nutrition/diet component. Other nutrient-related correlations link dietary fat and fiber to colon cancer, folate to the prevention of neural tube defects, calcium to the prevention of osteoporosis, psyllium to the lowering of blood lipid levels, and antioxidant nutrients to the scavenging of reactive oxidant species and protection against oxidative damage of cells that may lead to chronic disease. One group of phytochemicals, the isothiocyanates (glucosinolates, indoles, and sulforaphane), found in cruciferous vegetables such as broccoli, has been shown to trigger ensyme systems that block or suppress cellular DNA damage and that seem to reduce tumor size. The large numbers of phytochemicals that are implicated in this type of activity suggest that the potential impact of phytochemicals and functional foods on human and animal health is worth examining.

Beyond understanding of plant metabolism, the challenge then remains to better understand how components in the diet interact with human or animal metabolism to benefit their health and wellbeing. Although there exists extensive research and clinical support for specific nutrient effects as documented in the following sections, improving our knowledge at the fundamental level of molecular effects will better inform the decisions being made with respect to nutritional quality improvement. This challenge is at least as complex as the task of increasing or decreasing the amount of a specific protein, fatty acid, or other component of the plant itself. It is of little use producing a plant with a supposed nutritional benefit unless that benefit can be translated into positive health or nutritional impacts in humans or animals.

The application of rDNA technology to improve plant-specific components known to have benefit for human health that goes beyond meeting basic nutritional requirements is one way to introduce new functional foods into the marketplace. In addition to functional foods, rDNA technology allows the engineering of plants to address issues of animal nutrition and the impact of animal effluent on the environment. A good example of this is the addition of phytase enzymes to crops to reduce the need to add phosphorus to feed. Most of the phosphorus is added because the phosphorus in phytic acid is not bioavailable and because of the sequestering effect of phytic acid on uptake of divalent mineral ions.

The examples will illustrate the types of modifications that have been carried out or are being contemplated and describe their purpose, examine the successes and failures that have been documented, and provide insight into the technology used to produce nutritional alterations in plants so that readers will have a greater understanding of the problems that could arise from metabolic engineering.

Proteins and Amino Acids

Humans, as well as poultry, swine, and other nonruminant animals, have specific dietary requirements for each of the essential amino acids. A deficiency of essential amino acid limits growth and can be fatal.

In animal feeds, the primary limitations of maize and soybean meal-based diets are for lysine in nonruminant mammals and methionine in avian species. Maize with increased levels of lysine and soybeans with increased levels of methionine could allow diet formulations with improved amino acid balance, without the need to add crystalline lysine and methionine. Most plants have a poor balance of essential amino acids relative to the needs of animals and humans. The cereals (maize, wheat, rice, and so on) tend to be low in lysine, whereas legumes (soybean, peas, and so on) are often low in the sulphur-rich amino acids methionine and cysteine. Successful technical examples to date to enhance free amino acids levels include high-lysine maize and high-lysine canola and soybeans.

In areas such as less-developed countries, where it is difficult to obtain access to the components necessary for a balanced diet, these types of modifications could offer a particular advantage. Consumption of foods prepared from these crops potentially can help prevent protein malnutrition in such regions, especially among children, as well as increase the availability of animal protein in developing countries by improving the quality of animal feed. From an engineering perspective, one of the most straightforward methods to modify amino acid compositions of food and feed is by expressing proteins with high levels of the desired amino acids in the seed (the major food source).

One method of modifying storage protein composition is to introduce heterologous or homologous genes that code for proteins containing elevated levels of sulfur-containing amino acids (methionine, cysteine) and lysine. These proteins can be from other natural sources or can be synthetic. An example of the synthetic approach was published by Beauregard and others. An 11-kDa synthetic protein, MB1, was created to contain the maximum number of the essential amino acids methionine, threonine, lysine, and leucine in a stable, helical conformation. The structure was also designed to resist proteases to prevent degradation in-planta.

The high methionine (16%) and lysine (12%) contents make it a desirable candidate for improving soy protein quality. The MB1 protein was targeted to seed protein storage bodies using appropriate leader sequences and seed-specific promoters. Using a similar approach, another artificial storage protein (ASP-1) has been used to modify sweet potatoes. Transgenic plants exhibited a 2- and 5-fold increase in the total protein content in leaves and roots, respectively, over that of control plants. A significant increase in the level of essential amino acids such as methionine, threonine, tryptophan, isoleucine, and lysine was also observed.

The resulting tuber composition corresponds well with the World Health Organisation (WHO) standards for a nutritionally rich protein for optimal human nutrition. In this case, there was a striking increase in the growth rate and production of tubers in transgenic populations compared to the control. There was also an increase in the total protein content, with an increase in most essential amino acids. An important issue is that of ensuring that the total composition of storage proteins, for example, is not altered to the detriment of the development of the crop plant when attempting to improve amino acid ratios.

Rapp reported modifying soybean storage pro-teins in such a way that the 3-dimensional structure is maintained, and so that the modified proteins can accumulate in the seed at levels comparable to the endogenous seed proteins. A novel method of increasing essential amino acids was demonstrated by Lai and Messing. Maize produces a methionine-rich protein (delta-zein) in the grain but at a low level. Lai and Messing found a protein, Dzr1, that binds an intronic region and degrades delta-zein mRNA before translation. They replaced the targeted intronic region with an intron from another maize gene. This prevented Dzr1 from degrading delta-zein RNA and maximised the production of the methionine-rich protein. Chickens fed diets containing this maize grew significantly faster than chickens fed conventional maize.

This modification could potentially save animal farmers $1 billion per year in synthetic methionine supplements to maize-based feed. Attempts to manipulate the free lysine content of seeds illustrate that one needs to consider catabolic, as well as anabolic, variables when trying to engineer a particular metabolic phenotype in plants. A key step in lysine synthesis is catalysed by dihydrodipicolinate synthase (DHDPS), which is feedback inhibited by the pathway endproduct (lysine) and, thus, plays a key role in regulating flux through the pathway. Engineering plants to overexpress a feedback-insensitive bacterial DHDPS greatly increased flux through the lysine biosynthetic pathway.

However, in most cases this did not result in greater steady-state lysine levels because the plants also responded by increasing flux through the lysine catabolic pathway through elevation of lysine-ketoglutarate reductase. Substantial increases in lysine only occurred in plants where flux increased to such a level that the first ensyme of the catabolic pathway became saturated, again illustrating the potential complexities of metabolic regulation.

Carbohydrates

Plants make both polymeric carbohydrates (for example, starches and fructans) and individual sugars (for example, sucrose and fructose). The biosynthesis of these compounds is sufficiently understood to allow the bioengineering of their properties and to engineer crops to produce polysaccharides not normally present. The term prebiotic is used to describe an indigestible food ingredient, such as fructooligosaccharides (FOS), that beneficially affects the microflora by selectively stimulating the growth and/or activity of beneficial bacteria. Fructans (plant inulins) and fructooligosaccharides may be important ingredients in functional foods, because evidence suggests that they promote a healthy colon and help reduce the incidence of colon cancer.

The FOS may have anticarcinogenic, antimicrobial, hypolipidemic, and hypoglycemic actions in some. They may also help improve mineral absorption and balance, and may have antiosteoporotic and antiosteopenic activities. Inulins are only slightly digested in the small intestine. They are, however, fermented by a limited number of colonic bacteria. This could lead to changes in the colonic ecosystem in favor of some bacteria, such as Bifidobacteria, which may have health benefits.

Oral administration to humans of fructans, such as oligofructose and inulin, has been shown to increase the number of bifidobacteria in stools. Bifidobacteria may inhibit the growth of pathogenic bacteria, such as Clostridium perfringens and diarrheogenic strains of Escherichia coli. Inulins are considered to be bifidogenic factors. Their energy content is about half that of digestible carbohydrates or about 1 to 2 kcal/g. The possible anticarcinogenic activity might be accounted for, in part, by the possible anticarcinogenic action of butyrate.

Butyrate, along with other shortchain shortchain fatty acids, is produced by bacterial fermentation of FOS in the colon. Some studies have shown that butyrate induces growth arrest and cell differentiation and may also upregulate apoptosis, 3 activities that could be significant for antitumor activity. The FOS may lower serum triglyceride levels in some individuals. The mechanism of this possible effect is unclear. Decreased hepatocyte triglyceride synthesis is a hypothetical possibility. The FOS may also lower total cholesterol and LDL-cholesterol levels in some people. Again, the mechanism of this possible effect is unclear. Propionate, a product of FOS fermentation in the colon, may inhibit HMG-CoA reductase, the rate-limiting step in cholesterol synthesis.

Thus, there is interest in modifying plants to produce this polymeric carbohydrate. The main crop of interest for producing fructan is the sugar beet because the major storage component of this species is sucrose, the direct precursor for fructan biosynthesis. Seven-year and others have reported high-level fructan accumulation in a GM sugar beet without adverse effects on growth or phenotype. This work has implications both for the commercial manufacture of fructans and for the use of genetic engineering to obtain new products from existing crops.

Hellwege and others produced GM potato (Solanum tuberosum) tubers that synthesise the full spectrum of inulin molecules naturally occurring in globe artichoke (Cynara scolymus) roots. A similar approach (Allen and others 2002) is being used to derive soybean varieties that contain some oligofructan components that may selectively increase the population of beneficial species of bacteria (for example, Bifidobacteria) in the intestines of humans and certain animals and, thus, inhibit harmful species of bacteria (for example, E. Coli 0157:H7, Salmonella SE, and so on).

The soluble oligosaccharides, stachyose and raffinose, which are found in soybeans, are not digested and can cause flatulence and digestive problems, producing discomfort in humans. These compounds in conventional soybean or soybean meal are similarly not digested by nonruminant animals, resulting in reduced feed efficiency. Researchers found that the incorporation of lowstachyose soybean meal from nonmodified sources in prestarter pig diets tended to improve growth performance.

In addition, the increased sucrose content of lowstachyose soybean results in foods with a sweeter taste than do their traditional counterparts. Manipulating the level of this family of oligosaccharides through rDNA technology has been achieved by inhibiting galactinol synthase activity. This is the first committed step in the pathway and involves the synthesis of galactinol from UDP-Gal and myo-Inositol. The individual members are then synthesized by distinct galactose transferases (for example, raffinose synthase and stachyose synthase). As raffinose and stachyose may be crucial during seed development and storage, perhaps an alternate strategy would be that suggested by Griga and others, which is based on the transfer of α-galactosidase from a thermophilic bacterium (Thermotoga neapolitana) into legumes and inducing α-galactosidase to degrade the oligosaccharides after harvesting by changing the temperature. Starch is an important storage polysaccharide in many plants. It is composed of densely packed α-glucans, consisting of α-1,4-and α-1,6-linked glucose resides.

Engineering starch content and composition in potatoes is of interest. Plant ADP glucose pyrophosphorylase (ADPGPP) is sensitive to allosteric effectors and has been proposed to be a key regulator of starch biosynthesis. Stark and others engineered wild type and mutant allosterically insensitive E. Coli ADPGPP for chloroplast-targeted, tuberspecific expression in potatoes. Tubers from potato plants transformed with the allosterically insensitive E. Coli ADPGPP ensyme had starch levels up to 40% higher than the wild type. The higher starch content results in far less fat absorption during frying, because the moisture lost during frying is replaced by oil. However, there are still problems of irregular granule distribution throughout the tuber to be solved.

Schwall and others xcreated a potato producing very high amylose (slowly digested) starch by inhibiting ensymes that would normally make the amylopectin type of starch that is rapidly digested. This "resistant starch" is not digested in the small intestine, but is fermented in the large intestine by the microflora. Clinical studies have demonstrated that resistant starch has similar properties to fiber and has potential physiological benefits in humans.

Fiber and Lignans

Fiber is a group of substances chemically similar to carbohydrates, except that nonruminant animals poorly digest fiber. Fiber provides bulk in the diet, such that foods rich in fiber are satisfying without contributing significant calories. The technology to modify fiber content and type by genetic engineering would be a great benefit in persuading the many individuals who, for taste or other reasons, do not include adequate amounts of fiber in their daily diet. For example, fiber content could be added to more preferred foods or the more common sources of dietary fiber could be altered for greater health benefits. Other fiber-associated compounds include lignans. The 2 lignans of primary interest in mammals, enterodiol and its oxidation product, enterolactone, are formed in the intestinal tract by bacterial action on plant lignan precursors.

Flaxseed is the richest source of mammalian lignan precursors. Because enterodiol and enterolactone are structurally similar to both naturally occurring and synthetic estrogens, and have been shown to possess weakly estrogenic and antiestrogenic activities, they may play a role in the prevention of estrogen-dependent cancers. Genes encoding all the ensymes for the conversion of coniferyl alcohol (lignan and lignin precursor) to secoisolariciresinol, a major dietary phytoestrogen, have been cloned. Other alcohol derivatives such as plant sterols (mainly sitostanol) exhibit a dose-dependent action inhibiting cholesterol absorption while increasing cholesterol excretion and upregulating cholesterolgenesis in hamsters, resulting in lower circulating lipid levels.

However, as discussed elsewhere, low-fiber feedstuffs are often favored for nonruminant animals. Nonruminant animals do not produce ensymes necessary to digest cellulose-based plant fiber. Plants low in fiber should yield more digestible and metabolisable energy and protein and less manure and methane when fed to these species (North Carolina Cooperative Extension Service). US Dairy Forage Center (USDFRC) estimates that a 10% increase in fiber digestibility would result in an annual $350 million increase in milk/beef production and decreased generation of manure, manure, USDFRC estimates that a 10% increase in fiber digestibility is equivalent to 2.8 million tons decrease in manure solids produced each year.

Improved digestibility of livestock feed is therefore highly desirable. Guo and others developed low-lignin transgenic alfalfa through knockouts of enzymes involved in lignin biosynthesis. The altered lignin content and composition resulted in increased rate and extent of rumen digestion. Vermerris and Bout identified and cloned a brown midrib (Bmr) gene, which encodes caffeic acid O-methyltransferase (COMT), a lignin-producing ensyme. They generated mutants that give rise to plants that contain significantly lower lignin in their leaves and stems, leading to softer cell walls compared to wild type.

The plant-softening mutations improve the digestibility of the food, and livestock seem to prefer the taste. Such improved fiber digestibility in nonruminants should have significant beneficial effects because the efficiency of digestion of most highfiber diets for nonruminants is far from optimized.

Oils/lipids

Gene technology and plant breeding are combining to provide powerful means for modifying the composition of oilseeds to improve their nutritional value and provide the functional properties required for various food oil applications.

The technology also has the potential to produce industrial oils and chemicals in genetically engineered crops. Masur and others recently reviewed this topic. Genetic modification of oilseed crops can provide an abundant, relatively inexpensive source of dietary fatty acids with wide-ranging health benefits. Production of lipids shown to have health benefits in vegetable oil provides a convenient mechanism to deliver healthier products to consumers without the requirement for significant dietary changes. The lipid biosynthetic pathway was one of the earliest pathways to be targeted for modification, and it represents one of the better examples of metabolic engineering in plants to date.

Most ensymes required for fatty acid synthesis in plants have been cloned, and various academic and industrial groups have modified their expression to manipulate oilseed fatty acid composition. Major alterations in the proportions of individual fatty acids have been achieved in a range of oilseeds using conventional selection, induced mutation, and, more recently, posttranscriptional gene silencing. Examples of such modified oils include low- and sero-saturated fat soybean and canola oils, canola oil containing medium chain fatty acids (MCFA), high-stearic acid canola oil (for trans fatty acid-free products), high-oleic acid (monounsaturated) soybean oil, and canola oil containing the polyunsaturated fatty acids (PUFA), γ-linolenic (GLA; 18:3 n-6), stearidonic acids (SDA; C18:4 n-3), and other omega-3 fatty acids.

Altering the chain length and saturation level of the fatty acids can improve the nutritional qualities of some oils. In addition, genes from various plant species may be introduced to produce unusual fatty acids in oilseed crops. LauricalTM, canola oil with high amounts of lauric acid (C12:0), was the first commercial GM food oil. In this case, lauroyl-ACP thioesterase genes from the California bay laurel were cloned and transferred to canola (low-erucic acid rapeseed) oil crops. In 1995, the FDA completed its evaluation of Laurical for use in food products. Medium chain fatty acids (MCFA) range from 6 to 10 carbons long and are only minor components of natural foods.

The medium chain triglycerides (MCT) with these MCFA aid in absorption of calcium and magnesium and are rapidly oxidized as a quick source of energy. When MCT are substituted for long chain triglycerides (LCT) in the diet, animals gain less weight, store less adipose tissue, and experience an increase in metabolic rate. Mice fed diets with MCT have also been shown to possess increased endurance in swimming tests over that of mice fed diets with LCT.

Medium chain triglyceride oil has been included in medical foods, ergogenic aids, and dietary supplements. Because MCT are not readily available in high quantities in ordinary foods, they must be produced synthetically, making them of great interest to researchers. Thus, Dehesh and others have used the morilena mushroom and plants to identify ensymes involved in production of the MCT capric and caprylic acid. Expression of an acyl-ACP thioesterase cDNA from C. Hookeriana in seeds of canola, an oilseed crop that normally does not accumulate any capric and caprylic acid, resulted in a large increase in the levels of these 2 MCT. This illustrates the capacity to harness, through biotechnology, the genes contributing to phytochemical biodiversity in wild species and offers significant potential in the treatment of disease where such phytochemicals have proven health benefits.

Many types of fats are important, and the following sections will discuss different types of modifications with differing health implications. Edible oils rich in monounsaturated fatty acids provide improved oil stability, flavor, and nutrition for human and animal consumption. Oleic acid (C18:1), a monounsaturate, can provide more stability than the polyunsaturates, linoleic (C18:2) and linolenic (C18:3) acids. Higher monounsaturates are also preferred from a health perspective. Antisense inhibition of oleate desaturase expression in soybean resulted in oil that contained >80% oleic acid (23% is normal) and had a significant decrease in polyunsaturated fatty acids.

Clemente achieved a more stable effect using termination of transcripts with a self-cleaving ribozyme to enhance nuclear retention and serve as a tool to decrease specific plant gene expression achieving greater than 85% oleic, and saturated fatty acids levels of less than 6%. High-oleic soybean oil is naturally more resistant to degradation by heat and oxidation, and so requires little or no postrefining processing (hydrogenation), depending on the intended vegetable oil application. Liu and others produced high-stearic and high-oleic cottonseed oils by using posttranscriptional gene silencing.

While many lipids have important health implications, the longchain polyunsaturated fatty acids (PUFA), especially the omega-3 fatty acids found in fish, eicosapentaenoic acid (EPA) and docosahexaenoic acid (DHA), which are present in the retina of the eye and cerebral cortex of the brain, are some of the most well documented from a clinical perspective. Docosahexaenoic acid is also the predominant structural fatty acid in the gray matter of the brain. It is believed that EPA and DHA play an important role in the regulation of inflammatory immune reactions and blood pressure, treatment of conditions such as cardiovascular disease and cystic fibrosis, brain development in utero, and, in early postnatal life, the development of cognitive function

Omega-3 fatty acids also seem to be beneficial in certain neuropsychiatric illnesses such as bipolar disorder, schizophrenia, and depression. Current Western diets tend to be relatively high in omega-6 fatty acids and relatively low in omega-3 fatty acids. This is due in part to our high intake of vegetable oils that are rich in omega-6 fatty acids, and our low intake of oils and foods rich in omega-3 fatty acids, such as canola oil, flaxseed oil, or fatty fish. In plants, the microsomal ω-6 desaturase-catalysed pathway is the primary route of production of polyunsaturated lipids.

Ursin introduced genes encoding fatty acid desaturase from plants and fungi (such as the Δ-6 desaturase gene from a fungus (Mortierella) succeeding in producing omega-3 fatty acids in canola. In a clinical study designed to determine the relative efficacy of various fatty acids, metabolism of _-linolenic acid (ALA) and SDA, to the long-chain PUFA EPA, DPA n-3 (docosapentaenoic acid), and DHA in humans was measured. Researchers observed that SDA was superior in producing EPA by a factor of 3.6 over ALA.

Transgenic canola oil was obtained that contains >23% SDA, with an overall n-6:n-3 ratio of 0.5. Many food quality and health considerations encourage the development of oils containing altered ratios of saturated/unsaturated fatty acids.

Vitamins and Minerals

For selected minerals (iron, calcium, selenium, and iodine) and a limited number of vitamins (folate; vitamins E, B_6, and A), the clinical and epidemiological evidence is clear that they play a significant role in maintenance of optimal health and are limiting in diets worldwide. In addition, there is a growing knowledge base indicating that elevated intake of specific vitamins and minerals (for example, vitamins E and C, carotenoids, and selenium) may reduce the risk of diseases such as certain cancers, cardiovascular diseases, and chronic degenerative diseases associated with aging.

Because of the difficulty in separating individual nutrient effects from an overall dietary pattern that may be fundamental to achieving these health benefits, improved dietary patterns should still be encouraged. If nutrient intakes associated with optimal health benefits are not achievable by dietary modification alone, fortification of foods will be an alternative route. Genetic engineering is a potentially important route of fortification, particularly since it would seem to avoid many of the technical problems associated with food fortification such as uneven distribution of minute quantities of nutrients, unstable mixing and settling, over- or underaddition, and so on. Various groups (for example, the Consultative Group on International Agricultural Research) are using both traditional breeding and recombinant DNA approaches to develop biofortified crops that will be especially valuable in developing countries.

Rice is a staple that feeds nearly half the world's population, but milled rice does not contain β-carotene or significant amounts of its precursors. Integrating observations from prokaryotic systems into their work has enabled researchers to clone the majority of the carotenoid biosynthetic ensymes from plants during the 1990s. Ingo Potrykus and his research team at ETH-Zurich reported that immature rice endosperm is capable of synthesising the early intermediate of β-carotene biosynthesis.

Using carotenoid pathway genes from daffodil and Erwinia and a Rubisco transit peptide, his team succeeded in producing β-carotene in the rice endosperm. This major breakthrough in the modified rice plant led to the development of "Golden indica Rice" based on the concept reported earlier, which showed that an important step in provitamin A synthesis can be engineered into a non-green plant part that normally does not contain carotenoid pigments.

Chen and others took advantage of the fact that vitamin C can be scavenged by the ensyme dehydroascorbate reductase (DHAR) by introducing the gene encoding DHAR from wheat into maize and succeeded in increasing the amount of vitamin C by up to 100-fold. Iron is the most commonly deficient micronutrient in the human diet, and iron deficiency affects an estimated 1 to 2 billion people.

Anemia, characterised by low hemoglobin, is the most widely recognised symptom of iron deficiency, but there are other serious problems such as impaired learning ability in children, in-creased susceptibility to infection, and reduced work capacity. Three research groups led by Goto, Potrykus, and Datta employed the gene for ferritin, an iron-rich storage protein, under the control of an endosperm-specific promoter. Grain from these GM rice plants contained 3 times more iron than normal rice. To increase the iron content in the grain further, the researchers also focused on iron transport within the plant.

Nutraceuticals

The search for new compounds to treat human disease has led to the formation of specialised biotechnology firms searching for nutraceuticals (see the Glossary for a definition of the term nutraceutical). The recommended dietary allowances do not reflect the growing knowledge base, which indicates that elevated intakes of specific vitamins and minerals (that is, vitamins E and C, carotenoids, and selenium) significantly reduce the risk of diseases such as certain cancers, cardiovascular diseases, and chronic degenerative diseases associated with aging. To obtain such therapeutic levels in the diet, additional fortification of the food supply will be required as well as modification of dietary preferences, or direct modification of micronutrient levels in food crops. Studies by Bao and others and Bacon and others demonstrate that maximised dietary intake is not always correlated with optimised dietary benefit.

Quercetin is a flavonoid that has been demonstrated in some studies to work optimally at very low concentrations in protecting against cancerous cell proliferation and the actions of the carcinogen PhIP (2-

amino-1-methyl-6-phenylimidazo[4,5-b]pyridine) found in cooked meat. After activation in the liver, PhIP can attack DNA to form DNA adducts. Using accelerator mass spectrometry (AMS), this group has shown that both quercetin and sulforaphane can inhibit DNA adduct formation in a dose-dependent manner. The protective mechanism of quercetin is through the inhibition of the phase I enzyme CYP 1A2, while sulforaphane acts through the induction of phase II detoxification enzymes such as glutathi one transferases and UDP-glucuronosyl transferases. They further found that quercetin could ameliorate the effects of PhIP optimally at very low concentrations.

As the concentration was increased, the effect was attenuated. Similar effects may be found for other phytochemicals. This also illustrates the importance of taking a cautious approach to any research to increase phytochemicals with putative beneficial effects under the premise of "more is better." Unlike vitamins and minerals where mode of action is known, the primary evidence for the health-promoting roles of phytochemicals comes from epidemiological studies, and the exact chemical identity of many active compounds has yet to be determined. However, for select groups of phytochemicals, such as non-provitamin A carotenoids, glucosinolates, and phytoestrogens, the active compound or compounds have been identified and rigorously studied.

Other targets include improved iron content, through the production of iron-rich storage protein, bioavailable phosphorus released from phytate, and isoflavonoids.Other interesting products in the carotenoid pathway include lycopene, which may benefit the cardiovascular system by reducing the amount of oxidized low-density lipoprotein (LDL). Recent epidemiologic studies have suggested a potential benefit of this carotenoid in reducing the risk of prostate cancer, particularly the more lethal forms of this cancer.

Antinutrients

Reducing phytate is an example of a biotechnology approach that solves both a nutritional and an environmental problem. Seeds store the phosphorus needed for germination in the form of phytate, a sugar alcohol molecule having 6 phosphate groups (inositol hexaphosphate). However, phytate is an antinutrient because it strongly chelates iron, calcium, zinc, and other divalent mineral ions, making them unavailable for digestive uptake. Nonruminant animals generally lack the phytase enzyme needed for digestion of phytate. Poultry and swine producers in most countries currently add mined and processed (powdered) phosphate to the diets of their animals to enable optimal growth. Excess phosphate is excreted into the environment, resulting in water pollution.

When low-phytate soybean meal is utilized along with low-phytate maize for animal feeds, the phosphate excretion in swine and poultry manure is reduced by half. A series of GM rice lines (Japonica and Indica) have been developed to solve this problem. In addition, low-phytate maize was commercialized in the USA in 1999. Research indicates that the protein in low-phytate soybeans is also slightly more digestible than the protein in traditional soybeans. A number of studies have shown that optimal performance and bone mineralisation can result from diets without added phosphorus when phytase is included. Viveros and others demonstrated that phytase supplementation to low-phosphorus diets improved performance, mineral use, tibia weight, and relative liver weight in broiler chickens fed different levels of phosphorus.

Harper and others showed similar effects in growing-finishing swine. Phytase supplementation of low-phosphorus diets improves performance, phosphorus digestibility, and bone mineralisation and reduces phosphorus excretion in pigs. Poultry grew well on the engineered alfalfa diet without any inorganic phosphorus supplement. Thus, phosphorus supplements may be eliminated from poultry feed to reduce costs and reduce pollution. Other antinutrients that are being examined as possible targets for reduction are tryps in inhibitors, lectins, and several other heatstable components found in soybeans. Consideration must be given to possible increased susceptibility to pests and diseases when natural toxicants are removed, so the base germplasm should have input traits to counter this. Reducing the amounts of trypsin inhibitors in soybeans would have a positive effect on the domestic feed industry and offer a competitive advantage for on farm feeding of this protein source. If this can be combined with increases in the amounts of essential amino acids, very large improvements in productivity may be achieved.

While symptoms of food intolerance are common, true food allergy is less common. A food allergy is distinguished from food intolerance and other disorders by the production of antibodies (IgE) and the release of histamine and similar substances. The best-characterised true allergens include the superfamily cupins, which include globulins found in nuts and beans and albumins in nuts, and the superfamily prolamins found in cereal grains. Other common allergens are have in (initially from rubber trees), which causes contact dermatitis from latex, and chitinases.

Foods that frequently cause malabsorption or other food intolerance syndromes other than direct IgE immune responses include wheat and other gluten-containing grains (celiac disease or gluten-sensitive enteropathy is a multifactorial disorder caused by an inappropriate T-cell-mediated response to ingested gluten, resulting in chronic intestinal inflammation characterised by villous atrophy and malabsorption and cow's milk (milk/lactose intolerance and intolerance of dairy products-other than lactoglobulins, which are allergenic).

Allergen reduction by thioredoxin changes the biochemical and physical properties of proteins. According to present evidence, thioredoxin may be used to improve foods through, among other changes, lowering allergenicity and increasing digestibility. Using dogs, researchers have shown that thioredoxin reduces disulfide bonds of allergens (converting S-S to 2 SH), and thereby alters the allergenic properties of proteins extracted from wheat flour.

By changing the levels of expression of the thioredoxin gene, scientists have been able to reduce the allergenic effects of the protein fractions extracted from wheat and other cereals. Thioredoxin mitigated the allergenicity associated with the major protein fractions such as the gliadins (including the alpha, beta, and gamma types) and the glutenins, but gave less consistent results with the minor fractions, the albumins and globulins.

One soybean storage protein (P34) accounts for 85% of IgE responses in soybean-sensitive individuals. Sense suppression (gene silencing), driven by a seed-specific β-conglycinin promoter, was used to eliminate the accumulation of P34 in transgenic soybeans, removing the principal source of food allergenicity in soybeans. Early results from human blood serum tests indicate that P34-specific IgE antibodies could not be detected in soybean-sensitive people fed the gene-silenced beans.

Toxins

Plants are not always benign and produce many phytochemicals to protect themselves from pests. Over years of breeding and selection, most of the genes involved in the production of noxious products have been eliminated from plants used as food and feed crops. Potatoes and tomatoes are members of the deadly nightshade family and can contain toxic glycoalkaloids (for example, solanine) that have been linked to spina bifida.

Lectins are toxic glycoproteins that have the ability to bind to carbohydrate-containing molecules on the epithelial cells of the intestinal mucosa, thus causing toxicity. They are also called hemaglutinnins, based on their ability to agglutinate red blood cells. Kidney beans contain phytohemagglutinin and are poisonous if under cooked.

A number of people die each year from cyanogenic glycosides from peach and apricot seeds and many become ill from the sodium channel binding of grayanotox in honey produced from the nectar of rhododendrons. It is conceivable that biotechnology approaches can be employed to down regulate or even eliminate the genes involved in the metabolic pathways for the production, accumulation, and/or activation of these toxins in plants.

For example, the solanine content of potato has already been reduced substantially using an antisense approach, and efforts are under way to reduce the level of the other major potato glycoalkaloid, chaconine. Work has also been done to reduce cynaogenic glycosides in cassava through expression of the cassava ensyme hydroxynitrile lyase (HNL) in the roots.

Application of Modern Biotechnology Nutrient

The issue of what constitutes a significant change in intake of nutrients was discussed in the report of the International Food Biotechnology Council (IFBC). For nutrients, it was recommended that if a food supplies less than 5% of the average daily need (intake) in an amount of the food typically consumed per day by the population in question, then the intake from that source can be regarded as nonsignificant. Similarly, it could be stated that if the intake of any inherent constituent from a food or feed derived from a GM crop were increased by 5% or less, that would not be considered a significant change.

The distinction between a nonsignificant and a significant change is judgmental. The determination of the significance of a change in the level of a nutrient will also vary depending on the nutritional importance of the food and the availability of the nutrient in the food supply of the population. Recommended dietary intakes can be or have been set for most nutrients. Since each nutrient has a unique role and function and is present at different levels in different foods, the potential impact of changes in the dietary content of nutrients must be assessed on a case-bycase basis. It should also be recognised that certain new varieties may be developed to achieve a particular nutritional purpose within a specific age or gender group.

Per capita methods include food availability estimates or food disappearance data, presumably food eaten. Although per capita methods provide a representative general population mean of food consumption, they cannot provide consumption estimates for specific segments of the population. Specific segments may include populations who consume greater amounts of particular foods, either as a function of age, health status, or choice. Food consumption survey methods vary in their design and collection of dietary intake data and can range from 24-h dietary recalls to multiple-day dietary records.

A core component of the safety assessment of foods derived from GM crops is the molecular characterisation of the introduced DNA. A primary purpose of this analysis is to establish that the integrity of the vector DNA has not been modified as a result of the transformation process. The molecular characterisation of GM plants is comprised of essentially 2 basic components (1) a comprehensive description of the genetic elements and constructs used for plant transformation, and (2) the description of those elements as integrated in the transgenic event of interest. Outlined below are the generic requirements for molecular characterisation applied in North America. It should be noted that regulatory requirements for molecular characterisation may be different for Europe (EEC), Japan (Ministry of Heath and Welfare), Australia and New Zealand (ANZFA), Argentina, and other countries.

The constructs and transformation method used to generate the GM plant must be described. This includes a detailed description of the transformation method (for example, Agrobacterium-mediated transformation or direct transformation by methods such as particle bombardment, electroporation, or PEG transformation of protoplasts). For Agrobacterium-mediated transformation, the strain designation of any Agrobacterium used during the transformation process and how the Ti plasmid based vector was disarmed should be described, as well as the process used to free the system of remaining Agrobacterium cells once transformation was complete. For direct DNA-based transformation systems, the information should include information on whether the system utilised a pathogenic organism or nucleic acid sequences from a pathogen; how such sequences, if present, were removed prior to transformation; and whether the transformation process involved the use of helper plasmids or a mixture of plasmids or carrier DNA.

A detailed physical map of the vector used for transformation, including as appropriate the location of restriction sites, should be supplied, noting those portions of the vector used as primers in PCR analysis or as probes in Southern analysis. In addition, a summary of all genetic components that comprise the vector, including coding regions and noncoding sequences of known function, should be supplied.

The data on coding regions should detail the size of the individual DNA elements; the location, order, and orientation of the elements in the vector; the source of each element; and their probable function (if any) in the plant. In addition, information indicating whether any of the donor organisms or derived genetic components are known to cause disease or injury to plants or other organisms or are known toxicants, allergens, pathogenic factors, or irritants is supplied. If there is a history of safe use of the donor organism(s) or components thereof, that is also taken into account.

With regard to coding sequences (open reading frames), significant DNA sequence alterations to the native gene that resulted in a change in the amino acid sequence must be described. If the modified amino acid sequence has not been previously published, the complete sequence (highlighting the modifications) is to be reported, while DNA sequence modifications that affect only a few amino acids can be described without providing the complete sequence. Modifications known or anticipated to result in posttranslational modifications or alterations to the structure or function of the gene product must be described.

The complete nucleotide sequence of the DNA that is transformed into the plant is not generally required. However, sufficient data must be provided to demonstrate that the nature and order of the genetic elements as they existed in the vector DNA used in the transformation process have not been substantially altered following introduction into the plant. This may include Southern blot analysis, analysis with appropriate PCR analysis, DNA sequencing, RT-PCR data, and characterisation of the protein product produced from the inserted DNA to demonstrate that the expected protein is expressed in the plant.

Data describing the number of gene copies inserted into the plant, including the integration of partial gene fragments should be provided. In the case of allopolyploid plants, information identifying which parental genome the transgenic DNA has inserted into may also be required. The pattern and stability of inheritance of the introduced DNA (and gene function) must be demonstrated for plants that are male or female fertile, or both. A variety of methods can be used to demonstrate this, such as retention of phenotype, immunoassays, PCR, or Southern hybridization.

For plants that are infertile or for which it is difficult to produce seed (such as vegetatively propagated male-sterile potatoes), data must be provided to demonstrate that the transgenic trait is stably maintained and expressed during vegetative propagation over a number of generations appropriate for the crop. DNA is an integral part of every plant cell and is rapidly degraded by normal digestive processes, leading a number of organizations to conclude that consumption of DNA, including DNA introduced into GM crops, is safe. To date, fragments of low-copy plant transgenes have not been detected in the tissues of animals that are typically consumed by humans.

As with foods and feeds derived from GM crops with improved agronomic traits, the safety of any protein(s) that may be expressed from the inserted DNA in nutritionally improved products derived from GM crops as a result of any genetic change must be established. The need for studies to support safety requires consideration on a case-by-case basis and depends, in part, on available knowledge about the function and biological activity of the protein, as well as any history of prior exposure.

Where appropriate, safety studies may include standard animal testing to evaluate toxicological effects or immunological studies and bioinformatic approaches necessary to assess potential allergenicity. This may require the isolation of the protein from the plant or the synthesis of the protein by other means such as by E. coli, in which case there is a need to demonstrate biochemical, structural, and functional equivalence between this test material with that found in the plant. In 1993, OECD formulated the concept of substantial equivalence equivalence as a starting point for the safety assessment of GM crops.

Bibliography

Chapell, J., "Biochemistry and molecular biology of the isoprenoid biosynthetic pathway in plants", *Annual Review of Plant Physiology and Plant Molecular Biology,* 1995.

Christenson, SD; Wu, WM; Spies, MA; Shen, B; Toney, M.D., "Kinetic analysis of the 4-methylideneimidazole-5-one-containing tyrosine aminomutase in enediyne antitumor antibiotic C-1027 biosynthesis", *Biochemistry,* 2003.

Dewick, P.M., "Medicinal Natural Products: a biosynthetic approach", John Wiley & Sons, New York, 1997.

Dey, P.M. and Harborne, J.B., "Plant Biochemistry", Academic Press, London, 1997.

Fischer, M; Haase, I; Feicht, R; Schramek, N; Kohler, P; Schieberle, P; Bacher, A., "Evolution of vitamin B-2 biosynthesis: riboflavin synthase of Arabidopsis thaliana and its inhibition by riboflavin" *Biol. Chem.,* 2005.

Gabrielsen, M; Rohdich, F; Eisenreich, W; Grawert, T; Hecht, S; Bacher, A; Hunter W.N, "Biosynthesis of isoprenoids - A bifunctional IspDF enzyme from Campylobacter jejuni", *Eur. J. Biochem,* 2004.

Gang, D.R., Wang, J., Dudareva, N., Hee Nam, K., Simon, J.E., Lewinsohn, E. and Pichersky, E., "An investigation of the storage and biosynthesis of phenylpropenes in sweet basil", *Plant Physiology,* 2001.

Katz, L., and Donadio, S., "Polyketde synthesis: prospects for hybrid antibiotics", *Annual Review of microbiology,* 1993.

Kaufman, P.B., Cseke, L.J., Warber, S., Duke, J.A., Brielmann, H.L., *Natural Products from plants,* CRC Press, London. 1999.

Ketchan, T.M., "Alkaloid biosynthesis – the basics for metabolic engineering of medicinal plants", *Plant Cell,* 1995.

Laupitz, R; Hecht, S; Amslinger, S; Zepeck, F; Kaiser, J; Richter, G; Schramek, N; Steinbacher, S; Huber, R; Arigoni, D; Bacher, A; Eisenreich, W; Rohdich, F., "Biochemical characterization of Bacillus subtilis type II isopentenyl diphosphate isomerase, and phylogenetic distribution of isoprenoid biosynthesis pathways", *Eur. J. Biochem.* 2004.

Lichtenthaler, Hartmut K., Mueller, Christian, Schwender, Joerg, Jomaa, Hassan, Zeidler, Johannes, "Specific inhibition of isoprenoid (carotenoid, chlorophyll and isoprene) biosynthesis in plants by fosmidomycin", *Plant Physiology,* 1999.

Seemann, M; Bui, BTS; Wolff, M; Miginlac-Maslow, M; Rohmer, M., "Isoprenoid biosynthesis in plant chloroplasts via the MEP pathway: Direct thylakoid/ferredoxin-dependent photoreduction of GcpE/IspG", *Febs letters,* 2006.

Tellez, M, Dayan, F, Romagni, J, Schrader, K and S. Duke, "Terpenoid-based defense in plants and other related organisms", *In Lipid Biotechnology,* Marcel Dekker, New York, 2002..

Zepeck, F; Grawert, T; Kaiser, J; Schramek, N; Eisenreich, W; Bacher, A; Rohdich, F., "Biosynthesis of isoprenoids", *Purification and properties of IspG protein from Escherichia coli, J. Org. Chem.,* 2005.

Index